MELAMINE AND OTHER PROBLEMATIC FOOD CARCINOGENS

Melamine and Other Problematic Food Carcinogens

Viroj Wiwanitkit

Nova Science Publishers, Inc.
New York

Copyright © 2009 by Nova Science Publishers, Inc.

All rights reserved. No part of this book may be reproduced, stored in a retrieval system or transmitted in any form or by any means: electronic, electrostatic, magnetic, tape, mechanical photocopying, recording or otherwise without the written permission of the Publisher.

For permission to use material from this book please contact us:
Telephone 631-231-7269; Fax 631-231-8175
Web Site: http://www.novapublishers.com

NOTICE TO THE READER

The Publisher has taken reasonable care in the preparation of this book, but makes no expressed or implied warranty of any kind and assumes no responsibility for any errors or omissions. No liability is assumed for incidental or consequential damages in connection with or arising out of information contained in this book. The Publisher shall not be liable for any special, consequential, or exemplary damages resulting, in whole or in part, from the readers' use of, or reliance upon, this material.

Independent verification should be sought for any data, advice or recommendations contained in this book. In addition, no responsibility is assumed by the publisher for any injury and/or damage to persons or property arising from any methods, products, instructions, ideas or otherwise contained in this publication.

This publication is designed to provide accurate and authoritative information with regard to the subject matter covered herein. It is sold with the clear understanding that the Publisher is not engaged in rendering legal or any other professional services. If legal or any other expert assistance is required, the services of a competent person should be sought. FROM A DECLARATION OF PARTICIPANTS JOINTLY ADOPTED BY A COMMITTEE OF THE AMERICAN BAR ASSOCIATION AND A COMMITTEE OF PUBLISHERS.

LIBRARY OF CONGRESS CATALOGING-IN-PUBLICATION DATA
Viroj Wiwanitkit.
  Melamine and other problematic food carcinogens / author, Viroj Wiwanitkit.
     p. ; cm.
  Includes bibliographical references and index.
  ISBN 978-1-60692-940-7 (softcover)
   1. Carcinogens. 2. Food--Toxicology. 3. Food additives--Carcinogenicity. I. Title.
   [DNLM: 1. Food Contamination. 2. Carcinogens--toxicity. 3. Food Additives--toxicity. 4. Triazines--toxicity. WA 701 V819m 2009]
   RC268.7.F6V57   2009
   363.8'64--dc22
                                                                                          2008052989

Published by Nova Science Publishers, Inc. ✢ *New York*

Contents

Preface		vii
Chapter 1	Introduction to Nutritional Problem	1
Chapter 2	Melamine: A Present Hot Issue	11
Chapter 3	Fungal Contamination and Aflatoxin	21
Chapter 4	Grilled and Smoked Food	35
Chapter 5	High Lipid Food and Cancer	49
Chapter 6	Liver Fluke and Cholangiocarcinoma	61
Chapter 7	Nitrosamine	75
Chapter 8	Salted and Fermented Food	81
Chapter 9	Food Additive: Dye and Favor	87
Chapter 10	Environmental Contaminants	93
Chapter 11	Alcoholic Beverage	99
Index		107

PREFACE

Food contaminantion and carcinogenesis are of great public interest. The present situation concerning melamine contamination in China has become a wakeup call to nations around the world. It is time to be extremely concerned with food safety. The data on food contamination and carcinogenesis is useful and must be systematized collected. This data can be a reference in medical and biological science. In this book, important information concerning food contamination relating to carcinogenesis is presented and discussed.

Chapter 1

INTRODUCTION TO NUTRITIONAL PROBLEM

Nutritional problem is an important group of medical disorders. Because human beings have to eat, food becomes necessary thing for life. Eating food is for getting nutrient. Basically, the five essential nutrients for human beings include carbohydrate, protein, lipid, vitamin and mineral. Carbohydrate is the major primary source of energy for human beings. Protein is for structuring and repairing of parts of human bodies. Fat is the secondary or supplementation source of energy, therefore, excessive fat consumption can bring fatty appearance. While the other two nutrients, vitamin and mineral, are considered as micronutrients, which are required at a very few amounts but cannot lack.

The nutritional problem can be seen in all countries around the world. The most problematic nutritional disorder is "undernutrition" or "nutritional deficiency". Nutritional deficiency is the significant disorder of developing countries [1 – 20]. The most vulnerable group to this problem is the pediatric populations who cannot feed themselves. The nutritional deficiency can be owing to any kinds of nutrients. This problem can be due to insufficient ingestion or other causes (such as parasitic infestation). On the other hand, the "overnutrition" can be seen, especially in developed countries [21 – 40]. This can reflex the fact that balance in nutrition is the best. Too much or too few is not acceptable.

However, there is also another kind of nutritional problem in medicine. This problem is not related to the quantity but quality. The contamination of food is this problem. Contamination means having unwanted additional particles. This can be germs or chemicals. Food contamination is the present problem around the world.

EFFECT OF FOOD CONTAMINATION

Effect of food contamination is presented in medicine. If there is no effect, there cannot be this nutritional problem. Basically, food contamination can affect both food and the ones who ingest it, the consumer.

A. Effect on Food

Contaminant is directly added into the food. Therefore, it must have effect on the food. At least, contaminant makes deviation in natural content of that food. It might have some additional biochemical reactions on that food. These can be the problem for the consumer, which will be further described.

B. Effect on Consumer

Consumer who eats contaminant in the food can have several effects.
Acute diarrhea is common. Most contaminated germs and toxins can induce diarrhea. Cholera, salmonellosis and shigellosis are the best examples for the bacterial contaminant induced diarrhea [41 – 44]. Parasitic contaminant can also cause diarrhea as well as chronic parasitic infestation. The ones who have impaired immune status such as those with acquired immunodeficiency syndrome (AIDS) can have more symptoms. In some scenarios, long term effects of contaminant can be seen. The situation of carcinogenesis is noted and this is the focus in this book.

C. Effect on Food Producer

This is not a direct effect but should be mentioned. In the present day, banning of contaminated product is practiced around the world. Indeed, when the rumors on food contamination occur, the quoted food products are usually banned by general population. For sure, this affects the selling target of the food producer. How to increase the responsibility of the food producer and to create food safety culture in mind is the main query in the present day.

D. Effect on General Population

This is also another indirect effect of food contamination. As already mentioned, when a case of contamination is detailed, fear is the general response of the general population. For example, after the occurrence of botulism outbreak in a tropical Asian country due to home canned bamboo shoots, the general population in that country avoided eating any bamboo products for several months. This is the common behavior responding to any outbreak in medicine.

HOW CAN FOOD CONTAMINANT CONTAMINATE TO FOOD ?

"How can food contaminant contaminate to food ?" is an interesting question because this is the root cause of food contamination. Indeed, there are several modes that contaminant can contaminate to the food.

A. Accidental Contamination

Accidental contamination is the most common mode that contaminant can contaminate to the food. This is usually owing to poor hygiene [41 – 44]. Without good hand washing, cook who is a carrier of infectious disease can cause contamination into the food. This condition is well known as feco-oral transmission. The focused contaminants are bacterial germs (*E.coli*, cholera, *Shigella* spp., *Salmonella* spp. and etc.), parasitic germs (*Cryptosporidium* spp., *Ascaris* spp. and etc.) or viral germs (poliovirus, enterovirus and etc.). Apart from germ, chemical can contaminate into food in poor food manufacturing process.

B. Intentional Contamination

Sometimes, contaminant is intentionally added to food. This might be for favor. Monosodium glutamate (MSG) is the good example. This might also be for good appearance of food. For example, nitrite is added into the pork to make it red and brings appetite to the consumer when see that contaminated pork. This is owing to the fact that appetite is strongly related to the general appearance of the food. How to promote the thought on nutritional usefulness rather than external appearance of the food is the important focus in public health. In the worst case, contaminant is intentionally added into the food by sticky and no responsible food producer to reduce the material cost in food production process [45 – 46]. The melamine contamination in the food in the present day is the best example for this case.

C. Internal Contamination

Sometimes contaminant is the result of other contaminant in the food. The new contamination is internally generated in the food. This is usually the toxin. Aflatoxin due to fungal contamination is a good example. Another good example is the botulinum toxin owing to *Clostidium* spp bacterial contaminant in poor preserved canned food.

D. Terrorism [47 - 56]

Terrorism is the most serious case of food contamination. Terrorism by adding contaminant into food can be defined as intentional contamination. The purpose of this practice is to kill which is totally different from general intentional contamination. Arsenic is used as poison for killing for a long time. Many famous assassinations in history are owing to terrorism by food contaminations [57 – 61].

FOOD CONTAMINATION AND CARCINOGENESIS

Food contaminant and carcinogenesis is of public interest [62 – 66]. The present situation of melamine contamination in China becomes a hit to the world society. It is the time to concern on food safety. The data on food contaminant and carcinogenesis is useful and should be systematized collected. These data can be the reference in medical and biological science. In this book, important items on food contaminant relating to carcinogenesis will be discussed and presented. The coverage on topics on hot topic in food contaminant and carcinogenesis is the hallmark of this book.

The chapters in this book cover several problematic contaminants including melamine, fungal contamination and aflatoxin, nitrosamine, liver fluke and cholangiocarcinoma, grilled and smoked food, high lipid food and cancer, food additive (dye and favor), salted and fermented food, environmental contaminants and alcoholic beverage. University students and academic staff can use this book as their good reference.

REFERENCES

[1] Khor GL. Food-based approaches to combat the double burden among the poor: challenges in the Asian context. *Asia Pac. J .Clin. Nutr.* 2008;17 Suppl 1:111-5.

[2] Population Crisis Committee PCC. Nutrition and population: study of three countries. *Profamilia.* 1988 Dec;4(13):35-46.

[3] Deleuze Ntandou Bouzitou G, Fayomi B, Delisle H. Child malnutrition and maternal overweight in same households in poor urban areas of Benin. *Sante.* 2005 Oct-Dec;15(4):263-70.

[4] Gopalan C. Current food and nutrition situation in south Asian and south-east Asian countries. B*iomed. Environ. Sci.* 1996 Sep;9(2-3):102-16.

[5] Struble MB, Aomari LL. Position of the American Dietetic Association: Addressing world hunger, malnutrition, and food insecurity. *J. Am. Diet. Assoc.* 2003 Aug;103(8):1046-57.

[6] Tontisirin K, Nantel G, Bhattacharjee L. Food-based strategies to meet the challenges of micronutrient malnutrition in the developing world. *Proc. Nutr. Soc.* 2002 May;61(2):243-50.

[7] Ayres WS, Mccalla AF. Rural development, agriculture, and food security. *Finance Dev.* 1996 Dec:8-11.

[8] Mahmood S, Sheikh KH, Mahmood T. Food poverty and its causes in Pakistan. *Pak. Dev. Rev.* 1991 Winter;30(4 Pt 2):821-32.

[9] Population Crisis Committee PCC. Food and population: study of three countries. Profamilia. 1988 Dec;4(13):35-47.

[10] Ritchie JA. Towards better nutrition: lip service or a realistic fight? Nutr Health. 1986;4(2):113-23.

[11] Huddleston B. Confronting world hunger. CARE Briefs Develop Isssues. 1983;(3):1-8.

[12] Sanders TG. The problems of nutrition in Brazil. *Am. Univ. Field. Staff Rep. South Am.* 1982;(16):1-19.

[13] Solimano G, Hakim P. Nutrition and national development: the case of Chile. *Int. J. Health Serv.* 1979;9(3):495-510.

[14] Ravenholt A. Malnutrition in the Philippines. Am Univ Field Staff Rep Asia. 1982;(20):3-12.

[15] Krishnaswamy K. Perspectives on nutrition needs for the new millennium for South Asian regions. *Biomed. Environ. Sci.* 2001 Jun;14(1-2):66-74.

[16] Lipton M. Challenges to meet: food and nutrition security in the new millennium. *Proc. Nutr. Soc.* 2001 May;60(2):203-14.

[17] Teller CH, Culagovski M, del Canto J, Sáenz L, Aranda-Pastor J. Demographic dynamics in the food-nutrition problem: the search for effective strategies in Latin America. *Arch. Latinoam. Nutr.* 1982 Sep;32(3):663-81.
[18] Shetty P. Achieving the goal of halving global hunger by 2015. *Proc. Nutr. Soc.* 2006 Feb;65(1):7-18.
[19] James P. Marabou 2005: nutrition and human development. *Nutr. Rev.* 2006 May;64(5 Pt 2):S1-11.
[20] Johnecheck WA, Holland DE. Nutritional status in postconflict Afghanistan: evidence from the National Surveillance System Pilot and National Risk and Vulnerability Assessment. *Food Nutr. Bull.* 2007 Mar;28(1):3-17.
[21] Hainer V, Toplak H, Mitrakou A. Treatment modalities of obesity: what fits whom? *Diabetes Care.* 2008 Feb;31 Suppl 2:S269-77.
[22] Shrewsbury V, Wardle J. Socioeconomic status and adiposity in childhood: a systematic review of cross-sectional studies 1990-2005. *Obesity* (Silver Spring). 2008 Feb;16(2):275-84.
[23] Stocks T, Lindahl B, Stattin P. Healthy life style seems to reduce the risk of cancer. New support for the hypothesis that overweight and high blood glucose increase the cancer risk. Lakartidningen. 2007 Dec 19;104(51-52):3867-70.
[24] Chapman IM. Obesity in old age. *Front Horm. Res.* 2008;36:97-106.
[25] Bleich S, Cutler D, Murray C, Adams A. Why is the developed world obese? *Annu. Rev. Public Health.* 2008;29:273-95.
[26] Kumanyika SK. Global calorie counting: a fitting exercise for obese societies. *Annu. Rev. Public Health.* 2008;29:297-302.
[27] Duvigneaud N, Wijndaele K, Matton L, Philippaerts R, Lefevre J, Thomis M, Delecluse C, Duquet W. Dietary factors associated with obesity indicators and level of sports participation in Flemish adults: a cross-sectional study. *Nutr. J.* 2007 Sep 21;6:26.
[28] Lissner L, Troiano RP, Midthune D, Heitmann BL, Kipnis V, Subar AF, Potischman N. OPEN about obesity: recovery biomarkers, dietary reporting errors and BMI. *Int. J. Obes.* (Lond). 2007 Jun;31(6):956-61.
[29] Weker H. Simple obesity in children. A study on the role of nutritional factors. *Med. Wieku. Rozwoj.* 2006 Jan-Mar;10(1):3-191.
[30] Zalilah MS, Khor GL, Mirnalini K, Norimah AK, Ang M. Dietary intake, physical activity and energy expenditure of Malaysian adolescents. *Singapore Med. J.* 2006 Jun;47(6):491-8.

[31] Hill JO, Melanson EL. Overview of the determinants of overweight and obesity: current evidence and research issues. *Med. Sci. Sports Exerc.* 1999 Nov;31(11 Suppl):S515-21.

[32] Fang J, Wylie-Rosett J, Cohen HW, Kaplan RC, Alderman MH. Exercise, body mass index, caloric intake, and cardiovascular mortality. *Am. J. Prev. Med.* 2003 Nov;25(4):283-9.

[33] Kruger R, Kruger HS, Macintyre UE. The determinants of overweight and obesity among 10- to 15-year-old schoolchildren in the North West Province, South Africa - the THUSA BANA (Transition and Health during Urbanisation of South Africans; BANA, children) study. Public Health Nutr. 2006 May;9(3):351-8.

[34] Speakman JR. Obesity: the integrated roles of environment and genetics. *J. Nutr.* 2004 Aug;134(8 Suppl):2090S-2105S.

[35] Christiansen E, Garby L, Sørensen TI. Quantitative analysis of the energy requirements for development of obesity. *J. Theor. Biol.* 2005 May 7;234(1):99-106.

[36] Dehghan M, Akhtar-Danesh N, Merchant AT. Childhood obesity, prevalence and prevention. *Nutr. J.* 2005 Sep 2;4:24.

[37] Scali J, Siari S, Grosclaude P, Gerber M. Dietary and socio-economic factors associated with overweight and obesity in a southern French population. *Public Health Nutr.* 2004 Jun;7(4):513-22.

[38] Fiore H, Travis S, Whalen A, Auinger P, Ryan S. Potentially protective factors associated with healthful body mass index in adolescents with obese and nonobese parents: a secondary data analysis of the third national health and nutrition examination survey, 1988-1994. *J. Am. Diet. Assoc.* 2006 Jan;106(1):55-64.

[39] Stam-Moraga MC, Kolanowski J, Dramaix M, De Backer G, Kornitzer MD. Sociodemographic and nutritional determinants of obesity in Belgium. *Int. J. Obes. Relat. Metab. Disord.* 1999 Feb;23 Suppl 1:1-9.

[40] Stunkard AJ, Berkowitz RI, Schoeller D, Maislin G, Stallings VA. Predictors of body size in the first 2 y of life: a high-risk study of human obesity. *Int. J. Obes. Relat. Metab. Disord.* 2004 Apr;28(4):503-13.

[41] Stenberg A, Macdonald C, Hunter PR. How effective is good domestic kitchen hygiene at reducing diarrhoeal disease in developed countries? A systematic review and reanalysis of the UK IID study. *BMC Public Health.* 2008 Feb 22;8:71.

[42] Fewtrell L, Colford JM Jr. Water, sanitation and hygiene in developing countries: interventions and diarrhoea--a review. *Water Sci. Technol.* 2005;52(8):133-42.

[43] Oyemade A, Omokhodion FO, Olawuyi JF, Sridhar MK, Olaseha IO. Environmental and personal hygiene practices: risk factors for diarrhoea among children of Nigerian market women. *J. Diarrhoeal. Dis. Res.* 1998 Dec;16(4):241-7.

[44] Fewtrell L, Kaufmann RB, Kay D, Enanoria W, Haller L, Colford JM Jr. Water, sanitation, and hygiene interventions to reduce diarrhoea in less developed countries: a systematic review and meta-analysis. *Lancet. Infect. Dis.* 2005 Jan;5(1):42-52.

[45] Bhanti M, Taneja A. Contamination of vegetables of different seasons with organophosphorous pesticides and related health risk assessment in northern India. *Chemosphere.* 2007 Aug;69(1):63-8.

[46] Hinsz VB, Nickell GS, Park ES. The role of work habits in the motivation of food safety behaviors. *J. Exp. Psychol. Appl.* 2007 Jun;13(2):105-14.

[47] Miller RL, Israelsen C, Jensen J. Agroterrorism: a mixed methods study examining the attitudes and perceptions of Utah producers. J Agric Saf Health. 2008 Jul;14(3):273-82.

[48] Taylor MK. Food terrorism and food defense on the Web. *Med. Ref. Serv. Q.* 2008 Spring;27(1):81-96.

[49] Brandt AW, Sanderson MW, DeGroot BD, Thomson DU, Hollis LC. Biocontainment, biosecurity, and security practices in beef feedyards. *J. Am. Vet. Med. Assoc.* 2008 Jan 15;232(2):262-9.

[50] Stearns D. Intentional contamination: the legal risks and responsibilities. *J. Environ. Health.* 2008 Jan-Feb;70(6):58-9.

[51] Begley S. Weaponized hamburgers? Newsweek. 2007 Jul 16;150(3):49. Reforming the food safety system: what if consolidation isn't enough? *Harv. Law Rev.* 2007 Mar;120(5):1345-66.

[52] Crutchley TM, Rodgers JB, Whiteside HP Jr, Vanier M, Terndrup TE. Agroterrorism: where are we in the ongoing war on terrorism? *J. Food Prot.* 2007 Mar;70(3):791-804.

[53] Boisen CS. Title III of the Bioterrorism Act: sacrificing U.S. trade relations in the name of food security. *Am. Univ. Law Rev.* 2007 Jan;56(3):667-718.

[54] Yoon E, Shanklin CW. Food security practice in Kansas schools and health care facilities. J. Am Diet Assoc. 2007 Feb;107(2):325-329.

[55] Todd EC. Challenges to global surveillance of disease patterns. *Mar. Pollut. Bull.* 2006;53(10-12):569-78.

[56] Bossi P, Garin D, Guihot A, Gay F, Crance JM, Debord T, Autran B, Bricaire F. Bioterrorism: management of major biological agents. *Cell Mol. Life Sci.* 2006 Oct;63(19-20):2196-212.

[57] Fournier JH. Napoleon Bonaparte really was murdered. The weapon: rat poison. *Int. Surg.* 2007 Sep-Oct;92(5):308-10.
[58] Mari F, Polettini A, Lippi D, Bertol E. The mysterious death of Francesco I de' Medici and Bianca Cappello: an arsenic murder? *BMJ.* 2006 Dec 23;333(7582):1299-301.
[59] Lin X, Alber D, Henkelmann R. Elemental contents in Napoleon's hair cut before and after his death: did Napoleon die of arsenic poisoning? *Anal. Bioanal. Chem.* 2004 May;379(2):218-20.
[60] Allison B. Cause of death: the mystery surrounding the death of Napoleon. *Pharos Alpha Omega Alpha Honor Med. Soc.* 2002 Spring;65(2):16-9.
[61] Weider B, Fournier JH. Activation analyses of authenticated hairs of Napoleon Bonaparte confirm arsenic poisoning. *Am. J. Forensic Med. Pathol.* 1999 Dec;20(4):378-82.
[62] Wogan GN, Hecht SS, Felton JS, Conney AH, Loeb LA. Environmental and chemical carcinogenesis. *Semin. Cancer Biol.* 2004 Dec;14(6):473-86
[63] Belpomme D, Irigaray P, Hardell L, Clapp R, Montagnier L, Epstein S, Sasco AJ. The multitude and diversity of environmental carcinogens. *Environ. Res.* 2007 Nov;105(3):414-29
[64] Vainio H, Wilbourn J. Cancer etiology: agents causally associated with human cancer. *Pharmacol. Toxicol.* 1993;72 Suppl 1:4-11.
[65] Boffetta P. Human cancer from environmental pollutants: the epidemiological evidence. *Mutat. Res.* 2006 Sep 28;608(2):157-62.
[66] Ames BN, Gold LS. The causes and prevention of cancer: the role of environment. *Biotherapy.* 1998;11(2-3):205-20.

Chapter 2

MELAMINE: A PRESENT HOT ISSUE

INTRODUCTION [1 – 14]

At present, emerging problem of melamine contamination in milk products from East Asia is the global nutritional problem. In the outbreak of melamine intoxication, many infants who ingested contaminated milk became ill and died. The problem is owing to the high protein, inexpensive but poor quality, melamine contaminated, milk. Of interest, according to the nutritional surveillance, those contaminated products passed the screening for protein content at acceptable high level of protein. This is the false positive in laboratory analysis. It should be noted that the standard determination at present day is the determination of nitrogen content in the milk and further implies it as protein level. The trick in scenario of melamine contamination makes use of this weak point of the protein determination system. Because melamine also has high nitrogen content, it can falsely increase resulted protein content in the contaminated milk. This mimicks the fact. The melamine contaminated milk will seem like a good milk with high protein content. Melamine can increase nitrogen content upto 66.67 % which is equal to 16.66 % of protein. For sure, this has impact on marketing. Low cost but high protein milk is marvelous for the consumers.

Indeed, melamine is a compound that is widely used as a composition of many things in the present way. The well known kitchenware, melamineware is the best example. Cups and dishes are usually made of melamine in the present day. There vehicles are safe and cause no problem. However, melamine as contaminant in milk has many problems. This is totally different from the situation of using melamineware to hold food. The other applications of melamine are for production of fertilizer, fire exhauster, dish washing agent and etc.

Therefore, there have to be standard for contamination of melamine. The acceptable level of food contamination for melamin is below 0.015 ppm. The toxic level is about 2.5 ppm of human body weight. Concerning this value, the toxic level can be reached in an adult, average weight 70 kg, in case that that adult take upto 17.5 liters of melamine contaminated milk. This is the fact that why melamine contaminated milk could be undetectable in the market for a long time without any problem. Please note that melamine contaminated milk usually has its level stating that "not for infantile feeding". However, the problematic situation emerged when those contaminated milk products were introduced to infantile population. When melamine contaminated milk is used for infantile feeding, the infant, weight about 3 kg, will develop toxicity only if ingest only 0.75 liter of melamine contaminated milk. This is easier than the case of adult. In addition, the exact contamination level is significantly higher that the quoted non-safety lower limit. In the reports on present crisis, some problematic milk products had up to 65 ppm of contamination. Therefore, only a few infantile feeding can cause toxicity.

In addition to milk, the contamination of melamine can be seen in dairy products. Bread, cookie as well as cake are hit by this situation. Many reports on melamine contamination in bakery products emerge. This brings a great fear to general population. This topic on melamine contamination becomes the present focus of people around the world. This becomes public health hot issue at present.

Last but not least, secondary contamination into food owing to melamine can be seen. Although melamineware is considered safe for carrying food it can be used for only a specific certain condition. It is recommended that melamineware cannot be used for holding hot food because hot food can bring the secondary contamination of formaldehyde which can be internally generated in case that melamineware gets the heat upto 100 degree Celcius [15]. Sugita et al studied the relationship between the concentrations of formaldehyde and melamine released into 4% acetic acid from dishes and bowls made of melamine-formaldehyde resin [15]. According to this work,

the derived correlation between the concentrations of formaldehyde and melamine released at 95 degrees C was $y=0.4858x-0.2728$ ($r=0.8860$), where y is melamine concentration (ppm), x is formaldehyde concentration (ppm) and r is the correlation coefficient [15]. In addition, there are also many reports of migration of formaldehyde from melamineware exposed to acetic acid, a common composition of vinegar. Formaldehyde is the well known toxic substance that can also lead to cancer. Therefore, the strict practice to the recommendation in using melamineware is needed. This health education is necessary for general population. Also, it is the basic recommendation not to put the melamineware into the microwave apparatus. The reason is to prevent internal generation of formaldehyde as well. In addition, rough scratching on melamineware when cleaning it is not recommended. This can direct liberate the melamine content and this can cause contamination into the food when that problematic melamineware is used for food containing.

Table 1. Reports on formaldehyde and melamineware

Authors	Details
Sugita et al [15]	Sugita et al studied the relationship between the concentrations of formaldehyde and melamine released into 4% acetic acid from dishes and bowls made of melamine-formaldehyde resin [15].
Ishwata et al [16]	Ishwata et al studied migration of melamine and formaldehyde into food-simulating solvents from cups made of melamine resin under various conditions [16]. According to this work, migration of melamine from the cups being used at a cafeteria was detectable, but that of formaldehyde was undetectable when the cups were kept at 60 degrees C for 30 min with 4% acetic acid [16].
Lund and Petersen [17]	Lund and Petersen studied on migration of formaldehyde and melamine monomers from kitchen- and tableware made of melamine plastic [17].
Bradley et al [18]	Bradley et al performed a survey of the migration of melamine and formaldehyde from melamine food contact articles available on the UK market [18].
Smerasta et al [19]	Smerasta et al studied on migration and material tests of some food-contact plastic wares made in Thailand [19].
Sugita et al [20]	Sugita et al studied on release of formaldehyde and melamine from tableware made of melamine-formaldehyde resin [20].
Ishiwata [21]	Ishiwata reported on liquid chromatographic determination of melamine in beverages [21].

MELAMINE INTOXICATION

Melamine intoxication is the present emerging problem. There are many reports on this situation from China. Indeed, melamine cannot be absorbed via gastrointestinal tract into human body. It will excrete via kidney and can cause the problem. The most serious presentation of acute melamine intoxication is the acute nephrotoxicity which can result in acute renal failure and death. Until present, about 50,000 illnesses resulting in about 12,000 hospitalizations have occurred [22 -23]. However there are very few data on the nephrotoxic effects of melamine in humans [22 -23]. Previous researches on animal models have shown that melamine, in particular in combination with cyanuric acid, causes deposition and precipitation of birefringent crystals thereby causing renal failure [22 – 25]. According to the study of Reimschuesse et al [24], melamine and cyanuric acid appeared to have low toxicity when administered separately but they resulted in extensive renal crystal formation when administered together. Reimschuesse et al said that the renal failure due to this cause was similar to acute uric acid nephropathy in humans, in which crystal spherulites obstruct renal tubules [24]. In addition, Thompson et al noted that melamine-containing crystals posed distinct light microscopic, histochemical, and SEM/EDXA characteristics that helped their identification in tissue [25]. The important reports are summarized in Table 2.

Table 2. Reports on melamine intoxication

Authors	Details
Brown et al [22]	Brown et al reported on outbreaks of renal failure associated with melamine and cyanuric acid in dogs and cats in 2004 and 2007 [22].
Parry [11]	Parry reported the situation of contaminated infant formula sickens 6200 babies in China in BMJ [11].
Parry [9]	Parry said that China's tainted infant formula sickens nearly 13,000 infants [9]
Yang and Battle [23]	Yang and Battle reported on acute renal failure due to melamine intoxication [23].

MELAMINE AND CARCINOGENESIS

The carcinogenesis property of melamine is of interest. Because exposure to low level of contamination can be expected at more number than cases of acute melamine intoxication. There are many reports on this situation. Corresponding to its nephrotoxicity, melamine is strongly mentioned for induction of urinary bladder cancer. The important reports are summarized in Table 3.

Table 3. Reports on melamine and carcinogenesis

Authors	Details
Matsui-Yuasa et al [27]	Matsui-Yuasa et al mentioned for spermidine/spermine N1-acetyltransferase as a new biochemical marker for epithelial proliferation in rat bladder responding to melamine exposure [27].
Okumura et al [28]	Okumura et al reported on relationship between calculus formation and carcinogenesis in the urinary bladder of rats administered the non-genotoxic agents thymine or melamine [28].
Heck and Tyl [29]	Heck and Tyl reported on the induction of bladder stones by terephthalic acid, dimethyl terephthalate, and melamine (2,4,6-triamino-s-triazine) and its relevance to risk assessment [29].
Perrella and Boutwell [30]	Perrella and Boutwell mentioned for triethylenemelamine as an initiator of two-stage carcinogenesis in mouse skin which lacked the potential of a complete carcinogen [30].
National Toxicology Program [31]	This is a report on carcinogenesis bioassay of melamine in rats and mice [31].
Rao and Salaman [32]	Roe and Salaman reported studies on incomplete carcinogenesis of triethylene melamine (T.E.M.), 1,2-benzanthracene and beta-propiolactone, as initiators of skin tumour formation in the mouse [32].

To Response to the Problem on Melamine Contamination [1 – 14]

As a protocol, there is a need to set specific methods to control melamine contamination situation. In many countries, to avoid the problem of melamine contamination, the public health agency usually set a protocol that the food producers who import milk and dairy products from external for production of their goods have to take full responsibility on their whole process. They have to take regular surveillances on their materials and products. Safety from importing step is needed. The food producers have to make sure that their materials are free from melamine contamination. If the food producers feel not sure that there materials are free from melamine contamination, they have to stop production and send the suspected materials to have approval from free from melamine contamination. Confirmation process by governmental public agency has to be done and stopping of the production have to be continuous till the complete of approval process. If the result shows no contamination, further production can be done. On the other hand, in case that contamination is detected, stopping of the production has to be done. If not, that food producer must be legally acted. However, despite approval from the imported source or later confirmation on the suspected materials, the food produces still have to set the surveillance system on their products. This is the way to get the certification on food safety in general. The finalized survey is also useful to warrant the finalized product for free of melamine contamination before launched into the market. For the consumer, the selection of product must base on the quality. Checking of the label before buying is suggested. Product without label or with incomplete label should be avoided. The standard label has to clarify on the place of manufacturing, manufacturer, manufacturing date, expired date as well as the contents or ingredients of the product.

According to the Good Manufacturing Process (GMP), the food producers have to strictly use the food safety protocol. This must be applied from the staring point to the terminal. This is for assurance on their food product. Tracing back for any process should be probable. Conformation for international standards is the best aim. It is the promoting aim to guarantee no melamine contamination in the food product. For the governmental agency who takes the responsibility to this problem, the strict action to the problem is needed. Banning of detected contaminated food product must be strictly done. All products existing in the marketing should be kept away when the scenario of melamine contamination is approved. In addition, control of broadcasting on information dealing with safety of the food product by the food producers has to be strictly controlled.

CYROMAZINE [33 – 36]

Cyromazine is a derivative of melamine. This compound is widely used in production of insecticide and animal food. The cyromazine can be easily absorbed into plant and animal. Therefore, it can enter into the food chain of human beings easily. There is no evidence on the accurate count of contamination of cyromazine in eggs of farmed chicken fed with cyromazine contaminated animal food but the problem can be expected. This becomes a new concern. This is a new topic in egg safety after the recent outbreak of bird flu. The recent situation of contamination in East Asia brings attention from the world. The contamination in orange grown under the conditions of spraying of cyromazine based insecticide is also of interest. Contamination in orange has to be approved as well. The problem of cyromazie is the new emerging nutritional contamination of the world.

Similar to melamine, the GMP process is useful to face up the problem due to cyromazine contamination. Good control on animal farm and horticultural farm is needed. In poultry farm, the closed system is warranted similar to the case of bird flu outbreak. Control and surveillance for all processes are suggested. Strict practice and collaboration are the keys for succeed in fighting with the new emerging problem of cyromazine contamination in food.

REFERENCES

[1]   In brief: melamine. *Med. Lett. Drugs Ther.* 2008 Oct 20;50(1297):81-4.
[2]   Hung BP. China's tainted infant formula: What WHO should do. *BMJ.* 2008 Oct 13;337:a2073.
[3]   Lang L. FDA Updates Health Information Advisory on Melamine Contamination. *Gastroenterology.* 2008 Oct 8. [Epub ahead of print]
[4]   Lattupalli R, Yee J, Kolluru A. Nephrotoxicity of mala fide melamine: modern era milk scandal. *ScientificWorld Journal.* 2008 Oct 1;8:949-50.
[5]   Outbreak news. Melamine contamination, China. *Wkly Epidemiol. Rec.* 2008 Oct 3;83(40):358.
[6]   Coulombier D, Heppner C, Fabiansson S, Tarantola A, Cochet A, Kreidl P, Reintjes R. Melamine contamination of dairy products in China--public

health impact on citizens of the European Union. *Euro Surveill.* 2008 Oct 2;13(40). pii: 18998.
[7] Parry J. China's tainted milk scandal spreads around world. *BMJ.* 2008 Oct 1;337:a1890.
[8] Outbreak news. Melamine-contaminated powdered infant formula, China. *Wkly Epidemiol. Rec.* 2008 Sep 26;83(39):350.
[9] Parry J. China's tainted infant formula sickens nearly 13,000 babies. *BMJ.* 2008 Sep 24;337:a1802.
[10] Yagai S, Kubota S, Unoike K, Karatsu T, Kitamura A. Cyanurate-guided self-assembly of a melamine-capped oligo(p-phenylenevinylene). *Chem. Commun.*(Camb). 2008 Oct 7;(37):4466-8.
[11] Parry J. Contaminated infant formula sickens 6200 babies in China. *BMJ.* 2008 Sep 18;337:a1738.
[12] Cianciolo RE, Bischoff K, Ebel JG, Van Winkle TJ, Goldstein RE, Serfilippi LM. Clinicopathologic, histologic, and toxicologic findings in 70 cats inadvertently exposed to pet food contaminated with melamine and cyanuric acid. *J. Am. Vet. Med. Assoc.* 2008 Sep. 1;233(5):729-37.
[13] Lang L. FDA issues statement on diethylene glycol and melamine food contamination. *Gastroenterology.* 2007 Jul;133(1):5-6.
[14] Melamine adulterates component of pellet feeds. J Am Vet Med Assoc. 2007 Jul 1;231.
[15] Sugita T, Ishiwata H, Yoshihira K. Release of formaldehyde and melamine from tableware made of melamine-formaldehyde resin. *Food Addit. Contam.* 1990 Jan-Feb;7(1):21-7.
[16] Ishiwata H, Inoue T, Tanimura A. Migration of melamine and formaldehyde from tableware made of melamine resin. *Food Addit. Contam.* 1986 Jan-Mar;3(1):63-9.
[17] Lund KH, Petersen JH. Migration of formaldehyde and melamine monomers from kitchen- and tableware made of melamine plastic. *Food Addit. Contam.* 2006 Sep;23(9):948-55.
[18] Bradley EL, Boughtflower V, Smith TL, Speck DR, Castle L. Survey of the migration of melamine and formaldehyde from melamine food contact articles available on the UK market. *Food Addit Contam.* 2005 Jun;22(6):597-606.
[19] Smerasta P, Srivburuang P, Tongtan N, Ishiwata H, Yoshihira K. Migration and material tests of some food-contact plastic wares made in Thailand. *Eisei. Shikenjo Hokoku.* 1991;(109):105-6.

[20] Sugita T, Ishiwata H, Yoshihira K. Release of formaldehyde and melamine from tableware made of melamine-formaldehyde resin. *Food Addit. Contam.* 1990 Jan-Feb;7(1):21-7.

[21] Ishiwata H, Inoue T, Yamazaki T, Yoshihira K. Liquid chromatographic determination of melamine in beverages. *J. Assoc. Off Anal. Chem.* 1987 May-Jun;70(3):457-60.

[22] Brown CA, Jeong KS, Poppenga RH, Puschner B, Miller DM, Ellis AE, Kang KI, Sum S, Cistola AM, Brown SA. Outbreaks of renal failure associated with melamine and cyanuric acid in dogs and cats in 2004 and 2007. *J. Vet. Diagn. Invest.* 2007 Sep;19(5):525-31.

[23] Yang VL, Battle D. Acute renal failure from adulteration of milk with melamine. *ScientificWorld Journal.* 2008 Oct 9;8:974-5.

[24] Reimschuessel R, Gieseker CM, Miller RA, Ward J, Boehmer J, Rummel N, Heller DN, Nochetto C, de Alwis GK, Bataller N, Andersen WC, Turnipseed SB, Karbiwnyk CM, Satzger RD, Crowe JB, Wilber NR, Reinhard MK, Roberts JF, Witkowski MR. Evaluation of the renal effects of experimental feeding of melamine and cyanuric acid to fish and pigs. *Am. J. Vet Res.* 2008 Sep;69(9):1217-28.

[25] Thompson ME, Lewin-Smith MR, Kalasinsky VF, Pizzolato KM, Fleetwood ML, McElhaney MR, Johnson TO. Characterization of melamine-containing and calcium oxalate crystals in three dogs with suspected pet food-induced nephrotoxicosis. *Vet.Pathol.* 2008 May;45(3):417-26.

[26] Ogasawara H, Imaida K, Ishiwata H, Toyoda K, Kawanishi T, Uneyama C, Hayashi S, Takahashi M, Hayashi Y. Urinary bladder carcinogenesis induced by melamine in F344 male rats: correlation between carcinogenicity and urolith formation. *Carcinogenesis.* 1995 Nov;16(11):2773-7.

[27] Matsui-Yuasa I, Otani S, Yano Y, Takada N, Shibata MA, Fukushima S. Spermidine/spermine N1-acetyltransferase, a new biochemical marker for epithelial proliferation in rat bladder. *Jpn. J. Cancer Res.* 1992 Oct;83(10):1037-40.

[28] Okumura M, Hasegawa R, Shirai T, Ito M, Yamada S, Fukushima S. Relationship between calculus formation and carcinogenesis in the urinary bladder of rats administered the non-genotoxic agents thymine or melamine. *Carcinogenesis.* 1992 Jun;13(6):1043-5.

[29] Heck HD, Tyl RW. The induction of bladder stones by terephthalic acid, dimethyl terephthalate, and melamine (2,4,6-triamino-s-triazine) and its

relevance to risk assessment. *Regul Toxicol. Pharmacol.* 1985 Sep;5(3):294-313.
[30] Perrella FW, Boutwell RK. Triethylenemelamine: an initiator of two-stage carcinogenesis in mouse skin which lacks the potential of a complete carcinogen. *Cancer Lett.* 1983 Nov;21(1):37-41.
[31] National Toxicology Program. NTP Carcinogenesis Bioassay of Melamine (CAS No. 108-78-1) in F344/N Rats and B6C3F1 Mice (Feed Study). *Natl. Toxicol. Program. Tech Rep. Ser.* 1983 Mar;245:1-171.
[32] Roe FJ, Salaman MH. Further studies on incomplete carcinogenesis: triethylene melamine (T.E.M.), 1,2-benzanthracene and beta-propiolactone, as initiators of skin tumour formation in the mouse. *Br. J. Cancer.* 1955 Mar;9(1):177-203.
[33] Patakioutas G, Savvas D, Matakoulis C, Sakellarides T, Albanis T. Application and fate of Cyromazine in a closed-cycle hydroponic cultivation of bean (Phaseolus vulgaris L.). *J. Agric. Food Chem.* 2007 Nov 28;55(24):9928-35.
[34] Hernández-Borges J, Cifuentes A, García-Montelongo FJ, Rodríguez-Delgado MA. Combining solid-phase microextraction and on-line preconcentration-capillary electrophoresis for sensitive analysis of pesticides in foods. *Electrophoresis.* 2005 Feb;26(4-5):980-9.
[35] Hernández-Borges J, Rodríguez-Delgado MA, García-Montelongo FJ, Cifuentes A. Highly sensitive analysis of multiple pesticides in foods combining solid-phase microextraction, capillary electrophoresis-mass spectrometry, and chemometrics. *Electrophoresis.* 2004 Jul;25(13):2065-76.
[36] Marshall E. EPA regulators take on the Delaney clause. *Science.* 1984 May 25;224(4651):851-2.

Chapter 3

FUNGAL CONTAMINATION AND AFLATOXIN

FUNGAL CONTAMINATION IN FOOD

Fungus is an easily seen microbe. It can be seen everywhere in our environment. There are several kinds of fungus. Similar to other microbes, contamination of fungus in food can be expected. This is usually due to the problem of hygiene in keeping of food. Several interesting reports on fungal contamination in food are published and examples are hereby presented in Table 1.

Similar to other kinds of contaminants, fungal contamination in food can cause several health effects to the consumers.

Acute diarrhea can be seen though less common than seen in scenarios of other kinds of microbes. There are several outbreaks of acute diarrhea due to fungal contamination in food. Some important reports are listed in Table 2.

However, the cases of intoxication due to mycotoxin as an internal contamination owing to fungal contamination is more focused in medicine. There are many kinds of mycotoxins that can be problematic. Outbreaks of mycotoxin intoxication can be seen. There are many reports on outbreak in veterinarian science.

In 1997, Schneider et al reported on a field outbreak of in animals in the Republic of South Africa [24]. The affected sheep were characterized by haemorrhagic septicaemia, anaemia, leucocytopaenia and haemorrhagic tendencies [24]. By disease investigation process, toxigenic strains of *Stachybotrys chartarum* were isolated from the wheat and barley straw [24]. van Halderen et al also reported another field outbreak of chronic aflatoxicosis in dairy calves in the Republic of South Africa [25].

Table 1. Interesting reports on fungal contamination in food

Authors	Details
Coda et al [1]	Coda et al reported on long-term fungi inhibitory activity of water-soluble extract from Phaseolus vulgaris, cv Pinto, and sourdough lactic acid bacteria during bread storage [1].
Greenhill et al [2]	Greenhill et al reported on mycotoxins and toxigenic fungi in sago starch from Papua New Guinea [2].
Noonim et al [3]	Noonim et al reported on isolation, identification and toxigenic potential of ochratoxin A-producing *Aspergillus* species from coffee beans grown in two regions of Thailand [3].
Yabe et al [4]	Yabe et al reported on isolation of microorganisms and substances inhibitory to aflatoxin production [4]
Borutova et al [5]	Borutova et al reported on effects of deoxynivalenol and zearalenone on oxidative stress and blood phagocytic activity in broilers [5]
Keese et al [6]	Keese et al reported on ruminal fermentation patterns and parameters of the acid base metabolism in the urine as influenced by the proportion of concentrate in the ration of dairy cows with and without Fusarium toxin-contaminated triticale [6]
Dänicke et al [7]	Dänicke et al reported on effects of a Fusarium toxin-contaminated triticale, either untreated or treated with sodium metabisulphite (Na2S2O5, SBS), on weaned piglets with a special focus on liver function as determined by the 13C-methacetin breath test [7].
Maragos et al [8]	Maragos et al reported on extraction of aflatoxins B1 and G1 from maize by using aqueous sodium dodecyl sulfate [8].
Rouse and van Sinderen [9]	Rouse and van Sinderen reported on bioprotective potential of lactic acid bacteria in malting and brewing [9].
Watanabe [10]	Watanabe reported on production of mycotoxins by *Penicillium expansum* inoculated into apples [10].
Venturini et al [11]	Venturini et al reported on antimicrobial activity of extracts of edible wild and cultivated mushrooms against foodborne bacterial strains [11].

Authors	Details
Gregori et al [12]	Gregori et al reported on effects of potassium sorbate on postharvest brown rot of stone fruit [12].
Perni et al [13]	Perni et al reported on old atmospheric plasma disinfection of cut fruit surfaces contaminated with migrating microorganisms [13].
van Nierop et al [14]	van Nierop et al reported on optimised quantification of the antiyeast activity of different barley malts towards a lager brewing yeast strain [14].
Allen et al [15]	Allen et al reported on effect of ultrasonic treatment during cleaning on the microbiological condition of poultry transport crates [15].
Selwet [16]	Selwet reported on effect of organic acids on numbers of yeasts and mould fungi and aerobic stability in the silage of corn [16].
Nunez et al [17]	Nunez et al reported on effects of aging and heat treatment on whole yeast cells and yeast cell walls and on adsorption of ochratoxin A in a wine model system [17].
Di Cagno et al [18]	Di Cagno et al reported on use of selected sourdough strains of Lactobacillus for removing gluten and enhancing the nutritional properties of gluten-free bread [18].
Hamad [19]	Hamad reported on microbial spoilage of date rutab collected from the markets of Al-Hofuf city in the Kingdom of Saudi Arabia [19].
Bevilacqua et al [20]	Bevilacqua et al reported on metabiotic effects of *Fusarium* spp. on Escherichia coli O157:H7 and Listeria monocytogenes on raw portioned tomatoes [20].

For human beings, although there is no large outbreak of mycotoxin intoxication mycotoxin is still important food contaminant affecting human beings. Of several mycotoxin, aflatoxin is widely mentioned in medicine due to its impact on human health.

Table 2. Interesting reports on outbreak of diarrhea due to fungal contamination in food

Authors	Details
He et al [21]	He et al reported on an outbreak of poisoning from *Penicillium cyclopium* contaminated dried persimmon [21].
Muñoz et al [22]	Muñoz et al mentioned for *Saccharomyces cerevisiae* fungemia as an emerging infectious disease [22].
Koffi-Akoua et al [23]	Koffi-Akoua et al reported on cryptosporidium and candida in pediatric diarrhea in Abidjan [23].

AFLATOXIN

Aflatoxin is a mycotoxin. This toxin can be contaminated in food and causes several problems. Aflatoxin is internally generated by *Aspergillus flavus*, *A. parasiticus* and *A. nomius*. Of several fungal species, *A. flavus* is the most common problematic species. This fungus can be contaminated in beans and corns [26 – 28]. The acceptable level of aflatoxin in food is below 20 ppb [26 – 28]. The causative fungus can be well grown in food with moisture about 14-30% and atmosphere with moisture about 75% [26 – 28]. Therefore, this fungal contamination is common in tropical countries. The colony of the causative fungus can be seen by naïve eye [26 – 28]. It has yellowish green color and can be fluorescent under ultraviolet [26 – 28]. Of interest, aflatoxin, the mycotoxin, has very high resistance to heat [26 – 28]. It can insist to upto 260 degree Celcius [26 – 28]. Therefore, simple cooking cannot destroy this kind of food contaminant. There are many reports on contamination of aflatoxin in several foods. Important reports are hereby listed in Table 3.

According to Table 3, it can reflect the problems on the aflatoxin in food. The contaminated aflatoxin can induce several disorders. Aflatoxicosis is the well – known conditions responding to aflatoxin intoxication. Liver dysfunction and serious neurological alteration can be seen [40 – 47]. There are many interesting reports on outbreak of aflatoxcosis in the literatures.

Table 3. Interesting reports on aflatoxin contamination in food

Authors	Details
Guzmán-de-Peña and Peña-Cabriales [29]	Guzmán-de-Peña and Peña-Cabriales discussed on regulatory considerations of aflatoxin contamination of food in Mexico [29].
Keen and Martin [30]	Keen and Martin reported on evidence of aflatoxin carcinogenic in human beings in Swaziland [30]
Wood [31]	Wood reported on aflatoxins in domestic and imported foods and feeds [31].
Li et al [32]	Li et al reported on aflatoxins and fumonisins in corn from the high-incidence area for human hepatocellular carcinoma in Guangxi, China [32].
Torres Espinosa et al [33]	Torres Espinosa et al reported on quantification of aflatoxins in corn distributed in the city of Monterrey, Mexico [33].
Vesonder et al [34]	Vesonder et al reported on *Aspergillus flavus* and aflatoxins B1, B2, and M1 in corn associated with equine death [34].
Rojas et al [35]	Rojas et al reported on aflatoxins in newly harvested corn in Panama [35].
Abbas et al [36]	Abbas et al reported on aflatoxin and fumonisin contamination of commercial corn hybrids in Mississippi [36].
Razzaghi-Abyaneh et al [37]	Razzaghi-Abyaneh et al reported on a survey on distribution of Aspergillus section Flavi in corn field soils in Iran focusing on population patterns based on aflatoxins, cyclopiazonic acid and sclerotia production [37].
Amer [38]	Amer reported on aflatoxin contamination of developing corn kernels [38].
Waenlor and Wiwanitkit [39]	This paper presents a summary of reports of aflatoxin contamination in various foods and products, which have been carried out in Thailand between 1967-2001 [39]. According to this work, the accumulated data showed 38.9% of samples were highly contaminated with aflatoxin [39]. Over half of the contaminated samples were peanuts, milk, and poultry [39]. Waenlor and Wiwanitkit indicated that there was a significant difference between the type of food and seasonal influence, but not geographical influence [39].

Table 4. Interesting reports on outbreak of aflatoxicosis

Authors	Details
Probst et al r [40]	Probst et al reported outbreak of an acute aflatoxicosis in Kenya in 2004 [40].
Azziz-Baumgartner et al [41]	Azziz-Baumgartner et al reported on a case-control study of an acute aflatoxicosis outbreak, Kenya, 2004 [41].
Lewis et al [42]	Lewis et al reported on aflatoxin contamination of commercial maize products during an outbreak of acute aflatoxicosis in eastern and central Kenya [42].
Lye et al [43]	Lye et al reported on an outbreak of acute hepatic encephalopathy due to severe aflatoxicosis in Malaysia [43].
Chao et al [44]	Chao et al reported an outbreak of aflatoxicosis and boric acid poisoning in Malaysia [44].
Krishna et al [45]	Krishna et al reported on an outbreak of aflatoxicosis in Angora rabbits [45].
Robinson et al [46]	Robinson et al reported on waterfowl mortality caused by aflatoxicosis in Texas [46].
Greene et al [47]	Greene et al reported on disseminated intravascular coagulation complicating aflatoxicosis in dogs [47].

AFLATOXIN AND PRIMARY LIVER CANCER

An important long term effect of aflatoxin is the carcinogenesis. There are many evidences in medicine confirming that aflatoxin can be the cause of primary liver cancer. Guzmán de Peña reported on carcinogenic, mutagenic and toxic properties of these mycotoxins and their risk to humans and further discussed into the causal relationship between aflatoxins and hepatocellular carcinoma [48]. Guzmán de Peña said that aflatoxin B1-formamidopyrimidine was a determinant of the carcinogenic and mutagenic capabilities [48]. Groopman et al also said that aflatoxins could multiplicatively increase the risk of hepatic carcinoma in people chronically infected with hepatitis B virus, which illustrates the deleterious effect that even low toxin levels in the diet can pose for human health [49]. Basically, the toxicity of aflatoxin is similar to other insults. The primary effect is induction of hepatitis. This can be demonstrated by increased hepatic enzyme level, especially for SGOT and SGPT. This process is reversible. However, if the

Table 5. Interesting reports on aflatoxin and liver cancer

Authors	Details
Raoul [50]	Raoul discussed on natural history of hepatocellular carcinoma and current treatment options [50].
Wogan et al [51]	Wogan et al discussed on environmental and chemical carcinogenesis including aflatoxin [51].
McGlynn et al [52]	McGlynn et al reported on susceptibility to aflatoxin B1-related primary hepatocellular carcinoma in mice and humans [52].
Staib et al [53]	Staib et al discussed on TP53 and liver carcinogenesis [53].
Dominguez-Malagón and Gaytan-Graham [54]	Dominguez-Malagón and Gaytan-Graham discussed on hepatocellular carcinoma [54].
Kwak et al [55]	Kwak et al reported on role of phase 2 enzyme induction in chemoprotection by dithiolethiones [55].
Kensler et al[56]	Kensler et al reported on chemoprotection by organosulfur inducers of phase 2 enzymes: dithiolethiones and dithiins [56].
Montesano et al [57]	Montesano et al discussed on hepatocellular carcinoma covering from gene to public health [57].
Sherman [58]	Sherman discussed on hepatocellular carcinoma [58].
Li et al [59]	Li et al reported on aberrations of p53 gene in human hepatocellular carcinoma from China [59].
Vainio and Wilbourn [60]	Vainio and Wilbourn reported on cancer etiology focusing on agents causally associated with human cancer [60].
Groopman et al [61]	Groopman et al reported on molecular dosimetry of urinary aflatoxin-N7-guanine and serum aflatoxin-albumin adducts predicts chemoprotection by 1,2-dithiole-3-thione in rats [61].
Autrup et al [62]	Autrup et al reported on determination of exposure to aflatoxins among Danish workers in animal-feed production through the analysis of aflatoxin B1 adducts to serum albumin [62].
Bressac [63]	Bressac reported on selective G to T mutations of p53 gene in hepatocellular carcinoma from southern Africa [63].
Groopman and Kensler [64]	Groopman and Kensler reported on the use of monoclonal antibody affinity columns for assessing DNA damage and repair following exposure to aflatoxin B1 [64].
Vesselinovitch et al [65]	Vesselinovitch et al reported on neoplastic response of mouse tissues during perinatal age periods and its significance in chemical carcinogenesis [65].

pathology still exists, it will run into the second phase, the fatty liver, which is still reversible. In this phase, the abnormality, appearance of fatty infiltration, can be seen by ultrasonography. After this stage, all pathological disturbances will not be reversible. The next phase is cirrhosis then it progress to cancer. This is also concordant with the process of aflatoxin and primary liver caner. Indeed, there are many interesting reports on aflatoxin and liver cancer. The interesting publications will be shown in Table 5.

REFERENCES

[1] Coda R, Rizzello CG, Nigro F, De Angelis M, Arnault P, Gobbetti M. Long-term fungi inhibitory activity of water-soluble extract from Phaseolus vulgaris, cv Pinto, and sourdough lactic acid bacteria during bread storage. *Appl. Environ. Microbiol.* 2008 Oct 10.

[2] Greenhill AR, Blaney BJ, Shipton WA, Frisvad JC, Pue A, Warner JM. Mycotoxins and toxigenic fungi in sago starch from Papua New Guinea. *Lett. Appl. Microbiol.* 2008 Oct;47(4):342-7.

[3] Noonim P, Mahakarnchanakul W, Nielsen KF, Frisvad JC, Samson RA. Isolation, identification and toxigenic potential of ochratoxin A-producing Aspergillus species from coffee beans grown in two regions of Thailand. *Int J. Food Microbiol.* 2008 Aug 22. [Epub ahead of print]

[4] Yabe K, Yan PS, Song Y, Ichinomiya M, Nakagawa H, Shima Y, Ando Y, Sakuno E, Nakajima H. Isolation of microorganisms and substances inhibitory to aflatoxin production. *Food Addit Contam.* 2008 Sep;25(9):1111-7.

[5] Borutova R, Faix S, Placha I, Gresakova L, Cobanova K, Leng L. Effects of deoxynivalenol and zearalenone on oxidative stress and blood phagocytic activity in broilers. *Arch. Anim. Nutr.* 2008 Aug;62(4):303-12.

[6] Keese C, Meyer U, Rehage J, Spilke J, Boguhn J, Breves G, Dänicke S. Ruminal fermentation patterns and parameters of the acid base metabolism in the urine as influenced by the proportion of concentrate in the ration of dairy cows with and without Fusarium toxin-contaminated triticale. *Arch Anim. Nutr.* 2008 Aug;62(4):287-302.

[7] Dänicke S, Beineke A, Goyarts T, Valenta H, Beyer M, Humpf HU. Effects of a Fusarium toxin-contaminated triticale, either untreated or treated with sodium metabisulphite (Na2S2O5, SBS), on weaned piglets with a special focus on liver function as determined by the 13C-methacetin breath test. *Arch Anim Nutr.* 2008 Aug;62(4):263-86.

[8] Maragos CM. Extraction of aflatoxins B1 and G1 from maize by using aqueous sodium dodecyl sulfate. *J. AOAC Int.* 2008 Jul-Aug;91(4):762-7.

[9] Rouse S, van Sinderen D. Bioprotective potential of lactic acid bacteria in malting and brewing. *J. Food Prot.* 2008 Aug;71(8):1724-33.

[10] Watanabe M. Production of mycotoxins by Penicillium expansum inoculated into apples. *J. Food Prot.* 2008 Aug;71(8):1714-9.
[11] Venturini ME, Rivera CS, Gonzalez C, Blanco D. Antimicrobial activity of extracts of edible wild and cultivated mushrooms against foodborne bacterial strains. *J. Food Prot.* 2008 Aug;71(8):1701-6.
[12] Gregori R, Borsetti F, Neri F, Mari M, Bertolini P. Effects of potassium sorbate on postharvest brown rot of stone fruit. *J. Food Prot.* 2008 Aug;71(8):1626-31.
[13] Perni S, Shama G, Kong MG. Cold atmospheric plasma disinfection of cut fruit surfaces contaminated with migrating microorganisms. *J. Food Prot.* 2008 Aug;71(8):1619-25.
[14] van Nierop SN, Axcell BC, Cantrell IC, Rautenbach M. Optimised quantification of the antiyeast activity of different barley malts towards a lager brewing yeast strain. *Food Microbiol.* 2008 Oct;25(7):895-901.
[15] Allen VM, Whyte RT, Burton CH, Harris JA, Lovell RD, Atterbury RJ, Tinker DB. Effect of ultrasonic treatment during cleaning on the microbiological condition of poultry transport crates. *Br. Poult Sci.* 2008 Jul;49(4):423-8.
[16] Selwet M. Effect of organic acids on numbers of yeasts and mould fungi and aerobic stability in the silage of corn. *Pol. J. Vet Sci.* 2008;11(2):119-23.
[17] Nunez YP, Pueyo E, Carrascosa AV, Martínez-Rodríguez AJ. Effects of aging and heat treatment on whole yeast cells and yeast cell walls and on adsorption of ochratoxin A in a wine model system. *J. Food Prot.* 2008 Jul;71(7):1496-9.
[18] Di Cagno R, Rizzello CG, De Angelis M, Cassone A, Giuliani G, Benedusi A, Limitone A, Surico RF, Gobbetti M. Use of selected sourdough strains of Lactobacillus for removing gluten and enhancing the nutritional properties of gluten-free bread. *J Food Prot.* 2008 Jul;71(7):1491-5.
[19] Hamad SH. Microbial spoilage of date rutab collected from the markets of Al-Hofuf city in the Kingdom of Saudi Arabia. *J. Food Prot.* 2008 Jul;71(7):1406-11.
[20] Bevilacqua A, Cibelli F, Cardillo D, Altieri C, Sinigaglia M. Metabiotic effects of Fusarium spp. on Escherichia coli O157:H7 and Listeria monocytogenes on raw portioned tomatoes. *J. Food Prot.* 2008 Jul;71(7):1366-71.
[21] He S, Jiu Y, Bian H, Huang J, Ye S, Lan Z, Xin Y. An outbreak of poisoning from Penicillium cyclopium contaminated dried persimmon. *Biomed. Environ. Sci.* 1992 Jun;5(2):115-24.

[22] Muñoz P, Bouza E, Cuenca-Estrella M, Eiros JM, Pérez MJ, Sánchez-Somolinos M, Rincón C, Hortal J, Peláez T. Saccharomyces cerevisiae fungemia: an emerging infectious disease. *Clin. Infect. Dis.* 2005 Jun 1;40(11):1625-34.
[23] Koffi-Akoua G, Ferly-Therizol M, Kouassi-Beugre MT, Konan A, Timite AM, Assi Adou J, Assale G. Cryptosporidium and candida in pediatric diarrhea in Abidjan. Bull *Soc. Pathol. Exot Filiales.* 1989;82(4):451-7.
[24] Schneider DJ, Marasas WF, Dale Kuys JC, Kriek NP, Van Schalkwyk GC. A field outbreak of suspected stachybotryotoxicosis in sheep. *J. S. Afr. Vet. Assoc.* 1979 Jun;50(2):73-81.
[25] van Halderen A, Green JR, Marasas WF, Thiel PG, Stockenström S. A field outbreak of chronic aflatoxicosis in dairy calves in the western Cape Province. *J .S. Afr. Vet. Assoc.* 1989 Dec;60(4):210-1.
[26] Kurtzman CP, Horn B, Hessetine W. Aspergillus nomius: A New aflatoxin producing species related to A. flavus and A. tamarii. *Annie van Lecnw* 1987; 53: 147-158.
[27] Karunyavanich S. Toxins from some fungus. *J. Dept. Med. Sci.* 1972; 14;: 37.
[28] Carnaghan RBA, Crawford M. Relationship between ingestion of aflatoxin and primary liver cancer. *Brit Vet J.* 1963; 120: 201-204.
[29] Guzmán-de-Peña D, Peña-Cabriales JJ. Regulatory considerations of aflatoxin contamination of food in Mexico. *Rev. Latinoam. Microbiol.* 2005 Jul-Dec;47(3-4):160-4.
[30] Keen P, Martin P. Is aflatoxin carcinogenic in man? The evidence in Swaziland. Trop *Geogr Med.* 1971 Mar;23(1):44-53.
[31] Wood GE. Aflatoxins in domestic and imported foods and feeds. *J. Assoc. Off Anal. Chem.* 1989 Jul-Aug;72(4):543-8.
[32] Li FQ, Yoshizawa T, Kawamura O, Luo XY, Li YW. Aflatoxins and fumonisins in corn from the high-incidence area for human hepatocellular carcinoma in Guangxi, China. *J. Agric. Food Chem.* 2001 Aug;49(8):4122-6.
[33] Torres Espinosa E, Acuña Askar K, Naccha Torres LR, Montoya Olvera R, Castrellón Santa Anna JP. Quantification of aflatoxins in corn distributed in the city of Monterrey, Mexico. *Food Addit. Contam.* 1995 May-Jun;12(3):383-6.
[34] Vesonder R, Haliburton J, Stubblefield R, Gilmore W, Peterson S. Aspergillus flavus and aflatoxins B1, B2, and M1 in corn associated with equine death. *Arch Environ Contam. Toxicol.* 1991 Jan;20(1):151-3.

[35] Rojas V, Martin MC, Quinzada M. Aflatoxins in newly harvested corn in Panama. *Rev. Med Panama.* 2000;25:4-7.
[36] Abbas HK, Williams WP, Windham GL, Pringle HC 3rd, Xie W, Shier WT. Aflatoxin and fumonisin contamination of commercial corn (Zea mays) hybrids in Mississippi. *J. Agric. Food Chem.* 2002 Aug 28;50(18):5246-54.
[37] Razzaghi-Abyaneh M, Shams-Ghahfarokhi M, Allameh A, Kazeroon-Shiri A, Ranjbar-Bahadori S, Mirzahoseini H, Rezaee MB. A survey on distribution of Aspergillus section Flavi in corn field soils in Iran: population patterns based on aflatoxins, cyclopiazonic acid and sclerotia production. *Mycopathologia.* 2006 Mar;161(3):183-92.
[38] Amer MA. Aflatoxin contamination of developing corn kernels. *Commun. Agric. Appl. Biol. Sci.* 2005;70(3):281-93.
[39] Waenlor W, Wiwanitkit V. Aflatoxin contamination of food and food products in Thailand: An overview. *Southeast Asian J. Trop. Med. Public Health* 2003; 34 (Suppl 2): 184 – 90.
[40] Probst C, Njapau H, Cotty PJ. Outbreak of an acute aflatoxicosis in Kenya in 2004: identification of the causal agent. *Appl. Environ. Microbiol.* 2007 Apr;73(8):2762-4.
[41] Azziz-Baumgartner E, Lindblade K, Gieseker K, Rogers HS, Kieszak S, Njapau H, Schleicher R, McCoy LF, Misore A, DeCock K, Rubin C, Slutsker L; Aflatoxin Investigative Group. Case-control study of an acute aflatoxicosis outbreak, Kenya, 2004. *Environ Health Perspect.* 2005 Dec;113(12):1779-83.
[42] Lewis L, Onsongo M, Njapau H, Schurz-Rogers H, Luber G, Kieszak S, Nyamongo J, Backer L, Dahiye AM, Misore A, DeCock K, Rubin C; Kenya Aflatoxicosis Investigation Group. Aflatoxin contamination of commercial maize products during an outbreak of acute aflatoxicosis in eastern and central Kenya. *Environ Health Perspect.* 2005 Dec;113(12):1763-7.
[43] Lye MS, Ghazali AA, Mohan J, Alwin N, Nair RC. An outbreak of acute hepatic encephalopathy due to severe aflatoxicosis in Malaysia. *Am. J. Trop. Med Hyg.* 1995 Jul;53(1):68-72.
[44] Chao TC, Maxwell SM, Wong SY. An outbreak of aflatoxicosis and boric acid poisoning in Malaysia: a clinicopathological study. *J. Pathol.* 1991 Jul;164(3):225-33.
[45] Krishna L, Dawra RK, Vaid J, Gupta VK. An outbreak of aflatoxicosis in Angora rabbits. *Vet Hum. Toxicol.* 1991 Apr;33(2):159-61.
[46] Robinson RM, Ray AC, Reagor JC, Holland LA. Waterfowl mortality caused by aflatoxicosis in Texas. *J. Wildl Dis.* 1982 Jul;18(3):311-3.

[47] Greene CE, Barsanti JA, Jones BD. Disseminated intravascular coagulation complicating aflatoxicosis in dogs. *Cornell Vet.* 1977 Jan;67(1):29-49.
[48] Guzmán de Peña D. Exposure to aflatoxin B1 in experimental animals and its public health significance. *Salud Publica Mex.* 2007 May-Jun;49(3):227-35.
[49] Groopman JD, Kensler TW, Wild CP. Protective interventions to prevent aflatoxin-induced carcinogenesis in developing countries. *Annu Rev. Public Health.* 2008;29:187-203.
[50] Raoul JL. Natural history of hepatocellular carcinoma and current treatment options. *Semin. Nucl. Med.* 2008 Mar;38(2):S13-8.
[51] Wogan GN, Hecht SS, Felton JS, Conney AH, Loeb LA. Environmental and chemical carcinogenesis. Semin. Cancer Biol. 2004 Dec;14(6):473-86.
[52] McGlynn KA, Hunter K, LeVoyer T, Roush J, Wise P, Michielli RA, Shen FM, Evans AA, London WT, Buetow KH. Susceptibility to aflatoxin B1-related primary hepatocellular carcinoma in mice and humans. *Cancer Res.* 2003 Aug 1;63(15):4594-601.
[53] Staib F, Hussain SP, Hofseth LJ, Wang XW, Harris CC. TP53 and liver carcinogenesis. Hum Mutat. 2003 Mar;21(3):201-16.
[54] Dominguez-Malagón H, Gaytan-Graham S. Hepatocellular carcinoma: an update. *Ultrastruct Pathol.* 2001 Nov-Dec;25(6):497-516.
[55] Kwak MK, Egner PA, Dolan PM, Ramos-Gomez M, Groopman JD, Itoh K, Yamamoto M, Kensler TW. Role of phase 2 enzyme induction in chemoprotection by dithiolethiones. *Mutat Res.* 2001 Sep 1;480-481:305-15.
[56] Kensler TW, Curphey TJ, Maxiutenko Y, Roebuck BD. Chemoprotection by organosulfur inducers of phase 2 enzymes: dithiolethiones and dithiins. Drug Metabol *Drug. Interact.* 2000;17(1-4):3-22.
[57] Montesano R, Hainaut P, Wild CP. Hepatocellular carcinoma: from gene to public health. *J. Natl. Cancer Inst.* 1997 Dec 17;89(24):1844-51.
[58] Sherman M. Hepatocellular carcinoma. Gastroenterologist. 1995 Mar;3(1):55-66.
[59] Li D, Cao Y, He L, Wang NJ, Gu JR. Aberrations of p53 gene in human hepatocellular carcinoma from China. *Carcinogenesis.* 1993 Feb;14(2):169-73.
[60] Vainio H, Wilbourn J. Cancer etiology: agents causally associated with human cancer. *Pharmacol. Toxicol.* 1993;72 Suppl 1:4-11.
[61] Groopman JD, DeMatos P, Egner PA, Love-Hunt A, Kensler TW. Molecular dosimetry of urinary aflatoxin-N7-guanine and serum aflatoxin-

albumin adducts predicts chemoprotection by 1,2-dithiole-3-thione in rats. *Carcinogenesis.* 1992 Jan;13(1):101-6.

[62] Autrup JL, Schmidt J, Seremet T, Autrup H. Determination of exposure to aflatoxins among Danish workers in animal-feed production through the analysis of aflatoxin B1 adducts to serum albumin. *Scand. J. Work Environ. Health.* 1991 Dec;17(6):436-40.

[63] Bressac B, Kew M, Wands J, Ozturk M. Selective G to T mutations of p53 gene in hepatocellular carcinoma from southern Africa. *Nature.* 1991 Apr 4;350(6317):429-31.

[64] Groopman JD, Kensler TW. The use of monoclonal antibody affinity columns for assessing DNA damage and repair following exposure to aflatoxin B1. *Pharmacol. Ther.* 1987;34(2):321-34.

[65] Vesselinovitch SD, Rao KV, Mihailovich N. Neoplastic response of mouse tissues during perinatal age periods and its significance in chemical carcinogenesis. *Natl. Cancer Inst. Monogr.* 1979 May;(51):239-50.

Chapter 4

GRILLED AND SMOKED FOOD [1 – 13]

Heating is the way that brings food to be able to eat. Indeed, raw food can be eaten but can bring problem. Because the germ within the food can be existed, eating raw food can introduce to get infection. Bacterial and parasitic contamination is common in raw food and this can lead to severe infection, especially for diarrhea. Therefore, heat food before eat is the rule. There are several means to heat food. These included boiling, grilling, smoking and etc. Grilling is one of the most commonly used food heating. The method makes food to be heated and also bring good favor, smell as well as taste. However, despite good favor, smell and taste, there are some problematic contaminant in grilled food owing to the grilling process. Of interest, those contaminants are classified as carcinogens. For sure, this can be seen in smoked foods. Smoked food might have more problematic comparing to simple grilled food because smoked food has to exposure to smoke which is the main source of contaminants.

There are three important problematic contaminants due to grilling process. The first is the polycyclic aromatic hydrocarbon, which is due to heating of fatty substances. This is owing to the incomplete oxidation of fatty substances. This can be seen in many grilled foods. This substance is confirmed for the carcinogenic property. There are several reports on this substance and several cancers. Further details will be further discussed in specific heading on this substance. The second is the heterocyclic aromatic hydrocarbon. This is owing to the heating of amino acid at high temperature. There are upto 20 types of heterocyclic aromatic hydrocarbon. It is noted that 10 of 20 types of this substance are verified as carcinogens in animal models. Similar to polycyclic aromatic hydrocarbon, there are several reports on this substance and several cancers. Further details will be further discussed in specific heading on this substance. The last, third, is acrylamide. This substance is the result of reaction between a specific amino acid, asparagine, and sugar at high temperature (about 185 degree Celcius). Therefore, prolonged heating can make more increase in this substance. This substance has neurotoxic effect and is also reported for carcinogenic property in animal models. Similar to polycyclic aromatic hydrocarbon and heterocyclic aromatic hydrocarbon, there are several reports on this substance and several cancers. Further details will be further discussed in specific heading on this substance. It should be noted that aromatic hydrocarbons are the main problematic contaminant in grilled food. This is corresponding to the fact that the grilled food usually has good smell due to aromatic hydrocarbons but this smell is a fatal smell. Ones who eat the food with these contaminants can develop cancer and finally reach their deaths. Avoid the grilled food should be the common recommendation to general population.

Table 1. Reports on polycyclic aromatic hydrocarbon contamination in food and carcinogenesis

Authors	Details
Santarelli et al [14]	Santarelli et al reviewed on the topic processed meat and colorectal cancer [14].
Kuhn et al [15]	Kuhn et al reported on determination of polycyclic aromatic hydrocarbons in smoked pork by effect-directed bioassay with confirmation by chemical analysis [15].
Kobayashi et al [16]	Kobayashi et al reported on polycyclic aromatic hydrocarbons in edible grain in their a pilot study of agricultural crops as a human exposure pathway for environmental contaminants using wheat as a model crop [16].
Irigaray et al [17]	Irigaray et al reviewed on lifestyle-related factors and environmental agents causing cancer which included polycyclic aromatic hydrocarbon, which is contaminated in food, ingestion [17].
Belpomme et al [18]	Belpomme et al reported on the multitude and diversity of environmental carcinogens including polycyclic aromatic hydrocarbon [18].
Varlet et al [19]	Varlet et al reported on determination of polycyclic aromatic hydrocarbon profiles by GC-MS/MS in salmon processed by four cold-smoking techniques

Authors	Details
Reinik et al [20]	[19]. Reinik et al reported on polycyclic aromatic hydrocarbons in meat products and estimated polycyclic aromatic hydrocarbon intake by children and the general population in Estonia [20].
World Health Organization [21]	World Health Organization reported on safety evaluation of certain contaminants in food [21].
Perugini et al [22]	Perugini et al reported on polycyclic aromatic hydrocarbons in marine organisms from the Gulf of Naples, Tyrrhenian Sea [22].
Zanieri et al [23]	Zanieri et al reported on polycyclic aromatic hydrocarbons in human milk from Italian women: influence of cigarette smoking and residential area [23].
Llobet et al [24]	Llobet et al reported on exposure to polycyclic aromatic hydrocarbons through consumption of edible marine species in Catalonia, Spain [24].
van der Wielen et al [25]	van der Wielen et al reported on determination of the level of benzo(a)pyrene in fatty foods and food supplements [25].
Ciemniak [26]	Ciemniak reported on polycyclic aromatic hydrocarbons in herbs and fruit teas [26].
Jánská et al [27]	Jánská et al reported on optimization of the procedure for the determination of polycyclic aromatic hydrocarbons and their derivatives in fish tissue [27].
Pandey et al [28]	Pandey et al reported on induction of hepatic cytochrome P450 isozymes, benzo(a)pyrene metabolism and DNA binding following exposure to polycyclic aromatic hydrocarbon residues generated during repeated fish fried oil in rats [28].
Ibáñez et al [29]	Ibáñez et al reported on dietary intake of polycyclic aromatic hydrocarbons in a Spanish population [29].
García-Falcón and Simal-Gándara [30]	García-Falcón and Simal-Gándara reported on determination of polycyclic aromatic hydrocarbons in alcoholic drinks and the identification of their potential sources [30].
García-Falcón and Simal-Gándara [31]	García-Falcon and Simal-Gándara reported on polycyclic aromatic hydrocarbons in smoke from

Authors	Details
Ramesh [32]	different woods and their transfer during traditional smoking into chorizo sausages with collagen and tripe casings [31]. Ramesh reported on bioavailability and risk assessment of orally ingested polycyclic aromatic hydrocarbons [32].
Guillén and Sopelana [33]	Guillén and Sopelana reported on load of polycyclic aromatic hydrocarbons in edible vegetable oils: importance of alkylated derivatives [33].
Falcó et al [34]	Falcó et al reported on polycyclic aromatic hydrocarbons in foods and corresponding human exposure through the diet in Catalonia, Spain [34].
Pagliuca et al [35]	Pagliuca et al reported on determination of high molecular mass polycyclic aromatic hydrocarbons in a typical Italian smoked cheese by HPLC-FL [35].
Petito Boyce and Garry [36]	Petito Boyce and Garry reported on developing risk-based target concentrations for carcinogenic polycyclic aromatic hydrocarbon compounds assuming human consumption of aquatic biota [36].
Siegmund et al [37]	Siegmund et al reported on sensitive method for the determination of nitrated polycyclic aromatic hydrocarbons in the human diet [37]
Grova [38]	Grova et al reported on detection of polycyclic aromatic hydrocarbon levels in milk collected near potential contamination sources [38].
Mostafa [39]	Mostafa reported on monitoring of polycyclic aromatic hydrocarbons in seafoods from Lake Timsah [39].

POLYCYCLIC AROMATIC HYDROCARBON

As already mentioned, polycyclic aromatic hydrocarbon is the result of heating of fatty substances. Incomplete oxidation of fatty substances can generate this specific food contaminant. This can be seen in many grilled foods presenting with good smell and that good smell mimicks their danger. Polycyclic aromatic hydrocarbon is confirmed for the carcinogenic property. Eating a lot of this substance is relating to the development of breast cancer, lung cancer as well as gastric cancer. However, getting the smell contaminated with polycyclic aromatic hydrocarbon is more direct related to respiratory tract cancer. This can be seen in the situation of getting the polycyclic aromatic hydrocarbon due to tobacco smoke or environmental pollution due to traffic jam. As already mentioned, there are several papers on carcinogenesis due to polycyclic aromatic hydrocarbon. The important papers will be presented in Table 1.

HETEROYCLIC AROMATIC HYDROCARBON

As already mentioned, heterocyclic aromatic hydrocarbon is the result of heating of amino acids. High temperature heating of amino acid can generate this specific food contaminant. Similar to polycyclic aromatic hydrocardon, heterocyclic aromatic hydrocarbon can be seen in many grilled foods presenting with good smell and that good smell mimicks their danger. Heterocyclic aromatic hydrocarbon is confirmed for the carcinogenic property. Upto 10 kinds of this contaminant are confirmed for generation of cancer in animal models. The proposed cancers included liver cancer, intestinal cancer, skin cancer as well as breast cancer. The amount of contaminated heterocyclic aromatic hydrocarbon is owing to the period of cooking and temperature. The cooking with prolonged period and high temperature has more risk. Direct contact to flame is more risk. Therefore, boiling has lower risk than grilling. However, another important source for heterocyclic aromatic hydrocarbon is tobacco smoke, which might be more problematic than contamination in food. Also, it should be noted that the contaminated level in food is usually lower than the carcinogenic level in animal experiments. As already mentioned, there are several papers on carcinogenesis due to heterocyclic aromatic hydrocarbon. The important papers will be presented in Table 2.

ACRYLAMIDE

As already mentioned, acrylamide is the result of heating and complex formation between asparagines, an amino acid, and surgar. This is newly discovered substance. The cooking with prolonged period and high temperature can generate this problematic contaminant.

Table 2. Reports on heterocyclic aromatic hydrocarbon contamination in food and carcinogenesis

Authors	Details
Edenharder et al [40]	Edenharder et al reported on antimutagenic effects and possible mechanisms of action of vitamins and related compounds against genotoxic heterocyclic amines from cooked food [40].
Jolivette et al [41]	Jolivette et al reported on thietanium ion formation from the food mutagen 2-chloro-4-(methylthio)butanoic acid [41].
Williams and Iatropoulos [42]	Williams and Iatropoulos reported on inhibition by acetaminophen of intestinal cancer in rats induced by an aromatic amine similar to food mutagens [42].
McManus et al [43]	McManus et al reported on metabolism of 2-acetylaminofluorene and benzo(a)pyrene and activation of food-derived heterocyclic amine mutagens by human cytochromes P-450 [43].

Table 3. Reports on acrylamide contamination in food and carcinogenesis

Authors	Details
Doerge et al [44]	Doerge et al reviewed on using dietary exposure and physiologically based pharmacokinetic/pharmacodynamic modeling in human risk extrapolations for acrylamide toxicity [44].
Törnqvist et al [45]	Törnqvist et al reported on approach for cancer risk estimation of acrylamide in food on the basis of animal cancer tests and in vivo dosimetry [45].
Kütting et al [46]	Kütting et al reported on an association between self-reported acrylamide intake and hemoglobin adducts as biomarkers of exposure [46].
Parzefall [47]	Parzefall performed a minireview on the toxicity of dietary acrylamide [47].
Clement et al [48]	Clement et al reported on an expression profile of human cells in culture exposed to glycidamide, a reactive metabolite of the heat-induced food carcinogen acrylamide [48].

Authors	Details
Exon [49]	Exon performed a review of the toxicology of acrylamide [49].
Wirfält el [50]	Wirfält el reported on an associations between estimated acrylamide intakes, and hemoglobin AA adducts in a sample from the Malmö Diet and Cancer cohort [50].
Mojska et al [51]	Mojska et al reported on acrylamide content in potato crisps in Poland [51].
Boettcher et [52]	Boettcher et al reported on acrylamide exposure via the diet and focused on influence of fasting on urinary mercapturic acid metabolite excretion in humans [52].
Toda et al [53]	Toda et al reported on recent trends in evaluating risk associated with acrylamide in foods [53].
Petersen and Tran [54]	Petersen and Tran reported on exposure to acrylamide focusing on placing exposure in context [54].
Mucci and Adami [55]	Mucci and Adami reported on the role of epidemiology in understanding the relationship between dietary acrylamide and cancer risk in human beings [55].
Törnqvist [56]	Törnqvist reported on history of acrylamide in food [56].
Kütting et al [57]	Kütting et al reported on an influence of diet on exposure to acrylamide [57]..
Hagmar et al [58]	Hagmar et al reported on differences in hemoglobin adduct levels of acrylamide in the general population with respect to dietary intake, smoking habits and gender [58].
Rice [59]	Rice reported on the carcinogenicity of acrylamide [59].
Murkovic [60]	Murkovic reported on acrylamide in Austrian foods [60].
Rietjens and Alink [61]	Rietjens and Alink reported on toxic substances in food including acrylamide [61].
Granath and Törnqvist [62]	Granath and Törnqvist discussed whether acrylamide in food is hazardous to humans [62].
Dybing and Sanner [63]	Dybing and Sanner reported on risk assessment of acrylamide in foods [63].

Neurotoxicity of acrylamide in animal and human beings is confirmed. Other toxic effects of acrylamide include toxicity to reproductive system and carcinogenicity. In animal models, acrylamide is confirmed as a carcinogen for thyroid cancer, breast cancer, oral cancer, vaginal cancer, testicular cancer, lung cancer and skin cancer. However, the dosage of acrylamide that cause cancer is significant higher than normal contamination in food. As already mentioned, there are several papers on carcinogenesis due to acrylamide hydrocarbon. The important papers will be presented in Table 3.

REFERENCES

[1] Lampe JW. Diet, genetic polymorphisms, detoxification, and health risks.*Altern Ther. Health Med.* 2007 Mar-Apr;13(2):S108-11.
[2] Felton JS, Knize MG, Bennett LM, Malfatti MA, Colvin ME, Kulp KS. Impact of environmental exposures on the mutagenicity/carcinogenicity of heterocyclic amines. Toxicology. 2004 May 20;198(1-3):135-45.
[3] Turesky RJ. The role of genetic polymorphisms in metabolism of carcinogenic heterocyclic aromatic amines. *Curr .Drug Metab.* 2004 Apr;5(2):169-80.
[4] Strickland PT, Qian Z, Friesen MD, Rothman N, Sinha R. Metabolites of 2-amino-1-methyl-6-phenylimidazo(4,5-b)pyridine (PhIP) in human urine after consumption of charbroiled or fried beef. *Mutat Res.* 2002 Sep 30;506-507:163-73.
[5] Knize MG, Kulp KS, Salmon CP, Keating GA, Felton JS. Factors affecting human heterocyclic amine intake and the metabolism of PhIP. *Mutat Res.* 2002 Sep 30;506-507:153-62.
[6] Felton JS, Knize MG, Salmon CP, Malfatti MA, Kulp KS. Human exposure to heterocyclic amine food mutagens/carcinogens: relevance to breast cancer. *Environ. Mol. Mutagen.* 2002;39(2-3):112-8.
[7] Pfau W, Marquardt H. Cell transformation in vitro by food-derived heterocyclic amines Trp-P-1, Trp-P-2 and N(2)-OH-PhIP. *Toxicology.* 2001 Sep 14;166(1-2):25-30.
[8] Knize MG, Salmon CP, Pais P, Felton JS. Food heating and the formation of heterocyclic aromatic amine and polycyclic aromatic hydrocarbon mutagens/ carcinogens. *Adv. Exp. Med. Biol.* 1999;459:179-93.
[9] Vikse R, Reistad R, Steffensen IL, Paulsen JE, Nyholm SH, Alexander J. Heterocyclic amines in cooked meat. Tidsskr Nor Laegeforen. 1999 Jan 10;119(1):45-9.
[10] Gross GA, Fay L. Quantitative determination of heterocyclic amines in food products. *Princess Takamatsu Symp.* 1995;23:20-9.

[11] Grivas S. Synthetic routes to the food carcinogen 2 amino-3,8-dimethylimidazo[4,5-f]quinoxaline (8-MeIQx) and related compounds. *Princess Takamatsu Symp.* 1995;23:1-8.

[12] Sugimura T. Past, present, and future of mutagens in cooked foods. *Environ. Health Perspect.* 1986 Aug;67:5-10.

[13] Ohnishi Y, Kinouchi T, Tsutsui H, Uejima M, Nishifuji K. Mutagenic nitropyrenes in foods. *Princess Takamatsu Symp.* 1985;16:107-18.

[14] Santarelli RL, Pierre F, Corpet DE. Processed meat and colorectal cancer: a review of epidemiologic and experimental evidence. *Nutr. Cancer.* 2008 Mar-Apr;60(2):131-44.

[15] Kuhn K, Nowak B, Klein G, Behnke A, Seidel A, Lampen A. Determination of polycyclic aromatic hydrocarbons in smoked pork by effect-directed bioassay with confirmation by chemical analysis. *J. Food Prot.* 2008 May;71(5):993-9.

[16] Kobayashi R, Okamoto RA, Maddalena RL, Kado NY. Polycyclic aromatic hydrocarbons in edible grain: a pilot study of agricultural crops as a human exposure pathway for environmental contaminants using wheat as a model crop. *Environ. Res.* 2008 Jun;107(2):145-51.

[17] Irigaray P, Newby JA, Clapp R, Hardell L, Howard V, Montagnier L, Epstein S, Belpomme D. Lifestyle-related factors and environmental agents causing cancer: an overview. *Biomed. Pharmacother.* 2007 Dec;61(10):640-58.

[18] Belpomme D, Irigaray P, Hardell L, Clapp R, Montagnier L, Epstein S, Sasco AJ. The multitude and diversity of environmental carcinogens. *Environ. Res.* 2007 Nov;105(3):414-29.

[19] Varlet V, Serot T, Monteau F, Le Bizec B, Prost C. Determination of PAH profiles by GC-MS/MS in salmon processed by four cold-smoking techniques. *Food Addit. Contam.* 2007 Jul;24(7):744-57.

[20] Reinik M, Tamme T, Roasto M, Juhkam K, Tenno T, Kiis A. Polycyclic aromatic hydrocarbons (PAHs) in meat products and estimated PAH intake by children and the general population in Estonia. *Food Addit Contam.* 2007 Apr;24(4):429-37.

[21] World Health Organization, Geneva. Safety evaluation of certain contaminants in food. Prepared by the Sixty-fourth meeting of the Joint FAO/WHO Expert Committee on Food Additives (JECFA). *FAO Food Nutr. Pap.* 2006;82:1-778.

[22] Perugini M, Visciano P, Manera M, Turno G, Lucisano A, Amorena M. Polycyclic aromatic hydrocarbons in marine organisms from the Gulf of Naples, Tyrrhenian Sea. *J. Agric Food Chem.* 2007 Mar 7;55(5):2049-54.

[23] Zanieri L, Galvan P, Checchini L, Cincinelli A, Lepri L, Donzelli GP, Del Bubba M. Polycyclic aromatic hydrocarbons (PAHs) in human milk from Italian women: influence of cigarette smoking and residential area. *Chemosphere.* 2007 Apr;67(7):1265-74.
[24] Llobet JM, Falcó G, Bocio A, Domingo JL. Exposure to polycyclic aromatic hydrocarbons through consumption of edible marine species in Catalonia, Spain. *J. Food Prot.* 2006 Oct;69(10):2493-9.
[25] van der Wielen JC, Jansen JT, Martena MJ, De Groot HN, In't Veld PH. Determination of the level of benzo[a]pyrene in fatty foods and food supplements. *Food Addit Contam.* 2006 Jul;23(7):709-14.
[26] Ciemniak A. [Polycyclic aromatic hydrocarbons (PAHs) in herbs and fruit teas. Rocz Panstw Zakl Hig. 2005;56(4):317-22.
[27] Jánská M, Tomaniová M, Hajslová J, Kocourek V. Optimization of the procedure for the determination of polycyclic aromatic hydrocarbons and their derivatives in fish tissue: Estimation of measurements uncertainty. *Food Addit Contam.* 2006 Mar;23(3):309-25.
[28] Pandey MK, Yadav S, Parmar D, Das M. Induction of hepatic cytochrome P450 isozymes, benzo(a)pyrene metabolism and DNA binding following exposure to polycyclic aromatic hydrocarbon residues generated during repeated fish fried oil in rats. *Toxicol. Appl. Pharmacol.* 2006 Jun 1;213(2):126-34.
[29] Ibáñez R, Agudo A, Berenguer A, Jakszyn P, Tormo MJ, Sanchéz MJ, Quirós JR, Pera G, Navarro C, Martinez C, Larrañaga N, Dorronsoro M, Chirlaque MD, Barricarte A, Ardanaz E, Amiano P, Gonzálezi CA. Dietary intake of polycyclic aromatic hydrocarbons in a Spanish population. *J. Food Prot.* 2005 Oct;68(10):2190-5.
[30] García-Falcón MS, Simal-Gándara J. Determination of polycyclic aromatic hydrocarbons in alcoholic drinks and the identification of their potential sources. *Food Addit Contam.* 2005 Sep;22(9):791-7.
[31] García-Falcon MS, Simal-Gándara J. Polycyclic aromatic hydrocarbons in smoke from different woods and their transfer during traditional smoking into chorizo sausages with collagen and tripe casings. *Food Addit Contam.* 2005 Jan;22(1):1-8.
[32] Ramesh A, Walker SA, Hood DB, Guillén MD, Schneider K, Weyand EH. Bioavailability and risk assessment of orally ingested polycyclic aromatic hydrocarbons. *Int. J. Toxicol.* 2004;23(5):301-33.
[33] Guillén MD, Sopelana P. Load of polycyclic aromatic hydrocarbons in edible vegetable oils: importance of alkylated derivatives. *J. Food Prot.* 2004 Sep;67(9):1904-13.

[34] Falcó G, Domingo JL, Llobet JM, Teixidó A, Casas C, Müller L. Polycyclic aromatic hydrocarbons in foods: human exposure through the diet in Catalonia, Spain. *J. Food Prot.* 2003 Dec;66(12):2325-31.

[35] Pagliuca G, Gazzotti T, Zironi E, Serrazanetti GP, Mollica D, Rosmini R. Determination of high molecular mass polycyclic aromatic hydrocarbons in a typical Italian smoked cheese by HPLC-FL. *J. Agric. Food Chem.* 2003 Aug 13;51(17):5111-5.

[36] Petito Boyce C, Garry M. Developing risk-based target concentrations for carcinogenic polycyclic aromatic hydrocarbon compounds assuming human consumption of aquatic biota. *J. Toxicol Environ Health B Crit Rev.* 2003 Sep-Oct;6(5):497-520.

[37] Siegmund B, Weiss R, Pfannhauser W. Sensitive method for the determination of nitrated polycyclic aromatic hydrocarbons in the human diet. *Anal. Bioanal Chem.* 2003 Jan;375(1):175-81.

[38] Grova N, Feidt C, Crépineau C, Laurent C, Lafargue PE, Hachimi A, Rychen G. Detection of polycyclic aromatic hydrocarbon levels in milk collected near potential contamination sources. *J. Agric. Food Chem.* 2002 Jul 31;50(16):4640-2.

[39] Mostafa GA. Monitoring of polycyclic aromatic hydrocarbons in seafoods from Lake Timsah. *Int J. Environ. Health Res.* 2002 Mar;12(1):83-91.

[40] Edenharder R, Worf-Wandelburg A, Decker M, Platt KL. Antimutagenic effects and possible mechanisms of action of vitamins and related compounds against genotoxic heterocyclic amines from cooked food. *Mutat Res.* 1999 Jul 21;444(1):235-48.

[41] Jolivette LJ, Kende AS, Anders MW. Thietanium ion formation from the food mutagen 2-chloro-4-(methylthio)butanoic acid. *Chem. Res. Toxicol.* 1998 Jul;11(7):794-9.

[42] Williams GM, Iatropoulos MJ. Inhibition by acetaminophen of intestinal cancer in rats induced by an aromatic amine similar to food mutagens. *Eur. J. Cancer Prev.* 1997 Aug;6(4):357-62.

[43] McManus ME, Burgess WM, Veronese ME, Huggett A, Quattrochi LC, Tukey RH. Metabolism of 2-acetylaminofluorene and benzo(a)pyrene and activation of food-derived heterocyclic amine mutagens by human cytochromes P-450. *Cancer Res.* 1990 Jun 1;50(11):3367-76.

[44] Doerge DR, Young JF, Chen JJ, Dinovi MJ, Henry SH. Using dietary exposure and physiologically based pharmacokinetic/pharmacodynamic modeling in human risk extrapolations for acrylamide toxicity. *J. Agric Food Chem.* 2008 Aug 13;56(15):6031-8.

[45] Törnqvist M, Paulsson B, Vikström AC, Granath F. Approach for cancer risk estimation of acrylamide in food on the basis of animal cancer tests and in vivo dosimetry. *J. Agric Food Chem.* 2008 Aug 13;56(15):6004-12.
[46] Kütting B, Uter W, Drexler H. The association between self-reported acrylamide intake and hemoglobin adducts as biomarkers of exposure. *Cancer Causes Control.* 2008 Apr;19(3):273-81.
[47] Parzefall W. Minireview on the toxicity of dietary acrylamide. *Food Chem Toxicol.* 2008 Apr;46(4):1360-4.
[48] Clement FC, Dip R, Naegeli H. Expression profile of human cells in culture exposed to glycidamide, a reactive metabolite of the heat-induced food carcinogen acrylamide. *Toxicology.* 2007 Oct 30;240(1-2):111-24.
[49] Exon JH. A review of the toxicology of acrylamide. *J. Toxicol. Environ. Health B Crit. Rev.* 2006 Sep-Oct;9(5):397-412.
[50] Wirfält E, Paulsson B, Törnqvist M, Axmon A, Hagmar L. Associations between estimated acrylamide intakes, and hemoglobin AA adducts in a sample from the Malmö Diet and Cancer cohort. *Eur. J. Clin. Nutr.* 2008 Mar;62(3):314-23.
[51] Mojska H, Gielecińska I, Szponar L, Chajewska K. Acrylamide content in potato crisps in Poland. Rocz Panstw Zakl Hig. 2006;57(3):243-9.
[52] Boettcher MI, Bolt HM, Angerer J. Acrylamide exposure via the diet: influence of fasting on urinary mercapturic acid metabolite excretion in humans. *Arch Toxicol.* 2006 Dec;80(12):817-9.
[53] Toda M, Uneyama C, Yamamoto M, Morikawa K. Recent trends in evaluating risk associated with acrylamide in foods. --Focus on a new approach (MOE) to risk assessment by JECFA--.Kokuritsu Iyakuhin Shokuhin Eisei Kenkyusho Hokoku. 2005;(123):63-7.
[54] Petersen BJ, Tran N. Exposure to acrylamide: placing exposure in context. *Adv. Exp. Med. Biol.* 2005;561:63-76.
[55] Mucci LA, Adami HO. The role of epidemiology in understanding the relationship between dietary acrylamide and cancer risk in humans. *Adv. Exp. Med. Biol.* 2005;561:39-47.
[56] Törnqvist M. Acrylamide in food: the discovery and its implications: a historical perspective. *Adv. Exp. Med. Biol.* 2005;561:1-19.
[57] Kütting B, Schettgen T, Beckmann MW, Angerer J, Drexler H. Influence of diet on exposure to acrylamide--reflections on the validity of a questionnaire. *Ann. Nutr. Metab.* 2005 May-Jun;49(3):173-7.
[58] Hagmar L, Wirfält E, Paulsson B, Törnqvist M. Differences in hemoglobin adduct levels of acrylamide in the general population with respect to dietary intake, smoking habits and gender. *Mutat Res.* 2005 Feb 7;580(1-2):157-65.

[59] Rice JM. The carcinogenicity of acrylamide. *Mutat Res.* 2005 Feb 7;580(1-2):3-20.

[60] Murkovic M. Acrylamide in Austrian foods. *J. Biochem. Biophys. Methods.* 2004 Oct 29;61(1-2):161-7.

[61] Rietjens IM, Alink GM. Nutrition and health--toxic substances in food. Ned Tijdschr Geneeskd. 2003 Nov 29;147(48):2365-70.

[62] Granath F, Törnqvist M. Who knows whether acrylamide in food is hazardous to humans? *J. Natl. Cancer Inst.* 2003 Jun 18;95(12):842-3.

[63] Dybing E, Sanner T. Risk assessment of acrylamide in foods. *Toxicol. Sci.* 2003 Sep;75(1):7-15.

Chapter 5

HIGH LIPID FOOD AND CANCER

VITAMIN D RECEPTOR BSMI POLYMORPHISM, IS IT A RISK TO DEVELOP COLORECTAL ADENOMA ?

1. Introduction

It suggests that estrogen and, conceivably, nutritional phytoestrogens are protective substances against colorectal cancer for both sexes [1]. Prevention of colorectal, mammary, and prostate cancer may also relies on proper synthesis of the antimitotic prodifferentiating vitamin D hormonal metabolite 1,25-(OH)(2)-cholecalciferol (1,25-D3) or calcitriol [1]. Cytochrome-P450-hydroxylases responsible for synthesis (CYP27B1; 25-D3-1 alpha-hydroxylase) and catabolism (CYP24; 1,25-D3-24-hydroxylase) of 1,25-D3 are not only detectable in the renal but are also expressed in human colon cells, prostate cells, and mammary cells [1]. Basically, calcitriol is known widely for its effects on bone and mineral metabolism [2]. Cross et al said that that nutritional soy or genistein could optimize extrarenal 1,25-D3 synthesis, which could lead to growth control and in inhibition of tumor progression [1].

Epidemiological data suggest that low Vitamin D levels may have a significant role in the genesis of prostate cancer, colorectal carcinoma and somtimes other tumors [2]. In prostate, breast, colorectal, head and neck and lung carcinoma as well as lymphoma, leukemia and myeloma model systems calcitriol has important anti-tumor activity in vitro and in vivo [2]. Peterlik and Cross said that vitamin D deficits increased the risk of malignancies, particularly of colon, breast and prostate gland [3]. Calcitriol effects are related to an increase in G0/G1 arrest, induction of apoptosis and differentiation, modulation of expression of growth factor receptors [2]. Calcitriol potentiates the antitumor effects of several cytotoxic agents and inhibits motility and invasiveness of tumor cells and formation of new blood vessels [2]. The efficiency of vitamin D receptor-mediated intracellular signalling is stricted by the negative effects of hypovitaminosis D on extrarenal 25-hydroxyvitamin D-1alpha-hydroxylase activity and thus on the production of 1,25-dihydroxyvitamin D3 [3]. Attenuation of signal transduction from the ligand-activated vitamin D receptor and calcium-sensing receptor seems to be the major process by which calcium and vitamin D insufficiencies generate perturbation of cellular functions in bone, kidney, intestine, mammary and prostate glands, endocrine pancreas, vascular endothelium, and, significantly, in the immune system [3].

Recent studies have shown several polymorphisms to be detectable in the vitamin D receptor (VDR) gene, but the influence of VDR gene polymorphisms on VDR protein function and signaling is not largely known [4]. In addition, three adjacent restriction fragment length polymorphisms for BsmI, ApaI, and TaqI, respectively, at the 3' end of the VDR gene have been the most frequently studied [4]. The effect of vitamin D receptor polymorphism on the carcinoma risk is widely discussed [1 – 3]. For colorectal adenoma, there are only a few recent previous publications concerning the VDR, especially for BsmI [5 - 6]. In addition, the results are varies and the further metanalysis study is warranted. Here, the author performed a metanalytic study to summarize the knowledge on the pattern of VDR BsmI polymorphism and risk for colorectal adenoma.

2. Materials and Methods

A literature review to find the previous case-controls reports about the pattern of VDR BsmI polymorphism and risk for colorectal adenoma was performed. The author used the electronic search engine PubMed (www.pubmed.com) and Scopus (www.scopus.com) in searching for the literatures. The available reports were collected and extracted for the data about the pattern of VDR BsmI polymorphism. Those primary data were used for further metanalysis study.

Concerning the metanalysis study, the overall prevalence rate of each VDR BsmI polymorphism among overall cases and controls was calculated. Also, the association between prevalence rate and nationality of the populations was assessed using Chi square test. The SPSS 11.0 for Windows was used for statistical analysis in this study.

3. Results

According to the literature review, 2 reports [5 - 6] were recruited for further metanalysis (Table 1). According to the metanalysis, 570 cases and 634 controls were studied. The overall prevalence of wild (BB) type of VDR BsmI polymorphism among overall cases and controls are 21.6 % and 19.1 %. The subjects who have BB polymorphism have 1.08 times to develop colorectal adenoma comparing with the subjects who do not have BB polymorphism.

Table 1. Previous reports on the pattern of VDR BsmI polymorphism and risk for colorectal adenoma

Author	Nationality	Number of subjects (N)		Frequency of BsmI polymorphism (%)					
				Case			control		
		Case	Control	BB	Bb	bb	BB	Bb	bb
Kim et al [5]	USA	393	406	18.2	48.1	36.6	15.3	48.8	33.0
Boyapati et al [6]	USA	177	228	28.9	37.2	34.0	25.9	37.3	36.8

4. Discussion

The vitamin D endocrine system is mentioned for its involvement in a broad variety of biological processes including bone metabolism, modulation of the immune response, and regulation of cell proliferation and differentiation [4]. Variations in this endocrine system have, thus, been related to a lot of common diseases, including osteoarthritis (OA), diabetes, carcinoma, cardiovascular disease and tuberculosis [4]. Evidence to support this pleiotropic feature of vitamin D has included epidemiological reports on circulating vitamin D hormone levels, but also genetic epidemiological studies [4].

Colorectal carcinoma, classically revealed by a low digestive bleeding, which can be occult or exteriorized, is noted for possible correlation to the vitamin D metabolism. Many published epidemiological studies have shown an association between dietary factors, including calcium and vitamin D, and the incidence of colon carcinoma [7]. There is also recent evidence supporting a protective effect of calcium and vitamin D in the causative factor of colorectal neoplasia [8]. Recently the Calcium Polyp Prevention Study demonstrated that calcium supplementation can decrease the recurrence of colon polyps, but the effect depends on serum vitamin D levels [7]. The proposed mechanisms of direct action of vitamin D on colonic epithelium are regulation of growth factor and cytokine synthesis and signaling, as well as modulation of the cell cycle, apoptosis, and differentiation [9]. Vitamin D analogues with reduced hypercalcemic activity are also under clinical research for use against colon carcinoma and other cancers [9]. However, only a subset of patients responds to this therapy, most possibly owing to loss of VDR expression during tumour progression [9].

Many polymorphisms in the 3'-UTR region of the VDR gene can deviate transcriptional activity and mRNA stability in minigene reporter constructs [10]. One of these polymorphisms, BsmI, is located in intron 8 of the VDR gene [10]. It is thought that the 3'-UTR region of the VDR gene is linked in the regulation of mRNA stability, and, therefore, polymorphisms in this region are linked to the degradation of the VDR mRNA and consequently reduce receptor density [11]. Kim et al noted that having VDR BsmI BB polymorphism might have some probable protective effects for developing colorectal adenoma [5]. Here, the author studied the pattern of VDR BsmI polymorphism in the patients with colorectal adenoma comparing with controls. According to this study, having BB polymorphism affect only a little on developing colorectal adenoma.

RISK OF INGESTION OF HIGH LIPID CONTENT FOOD

As already mentioned in the first chapter, fat is an essential nutrient for all human beings. It is required to get this nutrient. Functionally, the lipid is the secondary source of energy adding to carbohydrate. This means lipid will be used only if the carbohydrate is not sufficient for production of energy for human body. Therefore, fat is usually deposited in adipose tissue and result in fatty appearance. Fat is a risk for many diseases. The widely mentioned condition is the cardiovascular system disorder [12 – 30]. However, there are also other conditions relating to the fatty body. Eating of high lipid content food is also mentioned for the relationship to the development of cancer. It is noted for the risk of ingestion of high lipid content food. Many cancers are related to this behavior. The proposed cancers include breast cancer, colon cancer as well as prostate cancer. Also, there are weak evidences for pancreas cancer, ovarian cancer and endometrial carcinoma. There are some epidemiological reports on the population with behavior of ingestion of high lipid content that show high incidence of breast cancer, colon cancer and prostate cancer. Higher incidence of cancer as well as death rate can be seen in the population with the risk behavior that those without. Indeed, fat is not only the source of energy for human body but also promoter for cancer. The generation of sexual hormone due to fat is believe to be a contributing factor to increased rate of breast cancer. The increase of bile production due to fat is believed to be a contributing factor to increased rate of colon cancer. In addition, fat is also related to the disturbance of cell membrane as well as the generation of free radicals. There are many interesting reports on high lipid content food and cancer. Interesting ones will be summarized and shown in Table 2.

Table 2. Reports on high lipid content food and carcinogenesis

Authors	Details
Moral et al [31]	Moral et al reported on high corn oil and high extra virgin olive oil diets have different effects on the expression of differentiation-related genes in experimental mammary tumors [31].
Myrthue et al [32]	Myrthue et al reported on the iroquois homeobox gene 5 is regulated by 1,25-dihydroxyvitamin D3 in human prostate cancer and regulates apoptosis and the cell cycle in LNCaP prostate cancer cells [32].
Chajès et al [33]	Chajès et al reported on association between serum trans-monounsaturated fatty acids and breast cancer risk in the E3N-EPIC Study [33].
Drăgan et al [34]	Drăgan et al reported on role of multi-component functional foods in the complex treatment of patients with advanced breast cancer [34].
Canadanović-Brunet et al [35]	Canadanović-Brunet et al reported on radical scavenging, antibacterial, and antiproliferative activities of *Melissa officinalis* L. extracts [35].
Colomer et al [36]	Colomer et al discussed on Giacomo Castelvetro's salads as anti-HER2 oncogene nutraceuticals since the 17th century [36].
Hallgren et al [37]	Hallgren et al reportd on apoptosis and tumor cell death in response to human alpha-lactalbumin made lethal to tumor cells [37].
Dogan et al [38]	Dogan et al reported on effects of high-fat diet and/or body weight on mammary tumor leptin and apoptosis signaling pathways in MMTV-TGF-alpha mice [38].

Authors	Details
Ashour [39]	Ashour reported on antibacterial, antifungal, and anticancer activities of volatile oils and extracts from stems, leaves, and flowers of *Eucalyptus sideroxylon* and *Eucalyptus torquata* [39].
Wei et al [40]	Wei et al reported on effects of different dietary fatty acid on expression of nuclear receptor genes in breast cancer of rats [40].
Pierce et al [41]	Pierce et al reported that telephone counseling helps maintain long-term adherence to a high-vegetable dietary pattern [41].
Escrich et al [42]	Escrich et al reported on molecular mechanisms of the effects of olive oil and other dietary lipids on cancer [42].
Allen et al [43]	Allen et al reported on phytanic acid: measurement of plasma concentrations by gas-liquid chromatography-mass spectrometry analysis and associations with diet and other plasma fatty acids [43].
Chen et al [44]	Chen et al reported on flaxseed alone or in combination with tamoxifen inhibits MCF-7 breast tumor growth in ovariectomized athymic mice with high circulating levels of estrogen [44].
Turck [45]	Turck reported on later effects of breastfeeding practice [45].
Drewnowski et al [46]	Drewnowski et al reported on genetic sensitivity to 6-n-propylthiouracil has no influence on dietary patterns, body mass indexes, or plasma lipid profiles of women [46].
Midthune et al [47]	Midthune et al mentioned for binary regression in truncated samples, with application to comparing dietary instruments in a large prospective study [47].
Djuric et al [48]	Djuric et al reported on effects of high fruit-vegetable and/or low-fat intervention on breast nipple aspirate fluid micronutrient levels [48].
Nagata et al [49]	Nagata et al reported on an association of maternal fat and alcohol intake with maternal and umbilical hormone levels and birth weight [49].
Ray et al [50]	Ray et al reported on diet-induced obesity and mammary tumor development in relation to estrogen

Authors	Details
Grainger [51]	receptor status [50]. Grainger reported on consumption of dietary supplements and over-the-counter and prescription medications in men participating in the Prostate Cancer Prevention Trial at an academic center [51].
Palmer et al [52]	Palmer et al reported on the impact of diet and micronutrient supplements on the expression of neuroendocrine markers in murine Lady transgenic prostate [52].
Narita et al [53]	Narita et al reported on candidate genes involved in enhanced growth of human prostate cancer under high fat feeding identified by microarray analysis [53].
Stacewicz-Sapuntzakis [54]	Stacewicz-Sapuntzakis reported on correlations of dietary patterns with prostate health [54].
Venkateswaran et al [55]	Venkateswaran et al reported on association of diet-induced hyperinsulinemia with accelerated growth of prostate cancer (LNCaP) xenografts [55].
Escrich et al [56]	Escrich et al reported on molecular mechanisms of the effects of olive oil and other dietary lipids on cancer [56].
Allen et al [57]	Allen et al discussed on measurement of plasma concentrations of phytanic acid by gas-liquid chromatography-mass spectrometry analysis and associations with diet and other plasma fatty acids [57].

REFERENCES

[1] Cross HS, Kallay E, Lechner D, Gerdenitsch W, Adlercreutz H, Armbrecht HJ. Phytoestrogens and vitamin D metabolism: a new concept for the prevention and therapy of colorectal, prostate, and mammary carcinomas. *J. Nutr.* 2004;134:1207S-1212S.

[2] Trump DL, Hershberger PA, Bernardi RJ, Ahmed S, Muindi J, Fakih M, Yu WD, Johnson CS. Anti-tumor activity of calcitriol: pre-clinical and clinical studies. *J. Steroid Biochem. Mol. Biol.* 2004;89-90:519-26.

[3] Peterlik M, Cross HS. Vitamin D and calcium deficits predispose for multiple chronic diseases. *Eur. J. Clin. Invest.* 2005;35:290-304.
[4] Uitterlinden AG, Fang Y, Van Meurs JB, Pols HA, Van Leeuwen JP. Genetics and biology of vitamin D receptor polymorphisms. *Gene.* 2004;338:143-56.
[5] Kim HS, Newcomb PA, Ulrich CM, Keener CL, Bigler J, Farin FM, Bostick RM, Potter JD. Vitamin D receptor polymorphism and the risk of colorectal adenomas: evidence of interaction with dietary vitamin D and calcium. *Cancer Epidemiol. Biomarkers Prev.* 2001;10:869-74.
[6] Boyapati SM, Bostick RM, McGlynn KA, Fina MF, Roufail WM, Geisinger KR, Wargovich M, Coker A, Hebert JR. Calcium, vitamin D, and risk for colorectal adenoma: dependency on vitamin D receptor BsmI polymorphism and nonsteroidal anti-inflammatory drug use? *Cancer Epidemiol. Biomarkers Prev.* 2003;12:631-7.
[7] Harris DM, Go VL. Vitamin D and colon carcinogenesis. J Nutr. 2004;134(12 Suppl):3463S-3471S.
[8] Martinez ME. Primary prevention of colorectal cancer: lifestyle, nutrition, exercise. *Recent Results Cancer Res.* 2005;166:177-211.
[9] Larriba MJ, Munoz A. SNAIL vs vitamin D receptor expression in colon cancer: therapeutics implications.*Br. J. Cancer.* 2005;92:985-9.
[10] Morrison NA, Qi JC, Tokita A, Kelly PJ, Crofts L, Nguyen TV, Sambrook PN, Eisman JA. Prediction of bone density from vitamin D receptor alleles. Nature. 1994; 367: 284-287.
[11] Beelman CA, Parker R. Degradation of mRNA in eukaryotes. Cell. 1995; 81: 179-183. Jesudason D, Wittert G. Endocannabinoid system in food intake and metabolic regulation. *Curr. Opin. Lipidol.* 2008 Aug;19(4):344-8.
[12] Cichosz G. Atherogenic properties of milk fat--facts or myths? Przegl Lek. 2007;64 Suppl 4:32-4.
[13] Bays H. Rationale for prescription omega-3-acid ethyl ester therapy for hypertriglyceridemia: a primer for clinicians. Drugs Today (Barc). 2008 Mar;44(3):205-46.
[14] Rudkowska I. Functional foods for cardiovascular disease in women. *Menopause Int.* 2008 Jun;14(2):63-9.
[15] Willett WC. Overview and perspective in human nutrition. *Asia Pac. J. Clin. Nutr.* 2008;17 Suppl 1:1-4.
[16] Layman DK, Clifton P, Gannon MC, Krauss RM, Nuttall FQ. Protein in optimal health: heart disease and type 2 diabetes. *Am. J. Clin. Nutr.* 2008 May;87(5):1571S-1575S.

[17] Dalainas I, Ioannou HP. The role of trans fatty acids in atherosclerosis, cardiovascular disease and infant development. *Int Angiol.* 2008 Apr;27(2):146-56.
[18] Lovegrove JA, Gitau R. Nutrigenetics and CVD: what does the future hold? *Proc. Nutr. Soc.* 2008 May;67(2):206-13.
[19] Clifton PM, Keogh J. Metabolic effects of high-protein diets. *Curr. Atheroscler. Rep.* 2007 Dec;9(6):472-8.
[20] Samaha FF, Foster GD, Makris AP. Low-carbohydrate diets, obesity, and metabolic risk factors for cardiovascular disease. *Curr .Atheroscler. Rep.* 2007 Dec;9(6):441-7.
[21] Damjanovic M, Barton M. Fat intake and cardiovascular response. *Curr. Hypertens Rep.* 2008 Feb;10(1):25-31.
[22] Chess DJ, Stanley WC. Role of diet and fuel overabundance in the development and progression of heart failure. *Cardiovasc. Res.* 2008 Jul 15;79(2):269-78.
[23] Astrup A, Dyerberg J, Selleck M, Stender S. Nutrition transition and its relationship to the development of obesity and related chronic diseases. *Obes. Rev.* 2008 Mar;9 Suppl 1:48-52.
[24] Theuwissen E, Mensink RP. Water-soluble dietary fibers and cardiovascular disease. *Physiol. Behav.* 2008 May 23;94(2):285-92.
[25] Sanal MG. The blind men 'see' the elephant-the many faces of fatty liver disease. *World J. Gastroenterol.* 2008 Feb 14;14(6):831-44.
[26] Dong F, Ren J. Fitness or fatness--the debate continues for the role of leptin in obesity-associated heart dysfunction. *Curr. Diabetes Rev.* 2007 Aug;3(3):159-64.
[27] Fitó M, de la Torre R, Farré-Albaladejo M, Khymenetz O, Marrugat J, Covas MI. Bioavailability and antioxidant effects of olive oil phenolic compounds in humans: a review. *Ann. Ist Super Sanita.* 2007;43(4):375-81.
[28] Pieters M, Vorster HH. Nutrition and hemostasis: a focus on urbanization in South Africa. *Mol .Nutr .Food Res.* 2008 Jan;52(1):164-72.
[29] Volk MG. An examination of the evidence supporting the association of dietary cholesterol and saturated fats with serum cholesterol and development of coronary heart disease. *Altern. Med. Rev.* 2007 Sep;12(3):228-45.
[30] Everitt AV, Hilmer SN, Brand-Miller JC, Jamieson HA, Truswell AS, Sharma AP, Mason RS, Morris BJ, Le Couteur DG. Dietary approaches that delay age-related diseases. *Clin. Interv. Aging.* 2006;1(1):11-31.
[31] Moral R, Solanas M, Garcia G, Grau L, Vela E, Escrich R, Escrich E. High corn oil and high extra virgin olive oil diets have different effects on the

expression of differentiation-related genes in experimental mammary tumors. *Oncol. Rep.* 2008 Aug;20(2):429-35.

[32] Myrthue A, Rademacher BL, Pittsenbarger J, Kutyba-Brooks B, Gantner M, Qian DZ, Beer TM. The iroquois homeobox gene 5 is regulated by 1,25-dihydroxyvitamin D3 in human prostate cancer and regulates apoptosis and the cell cycle in LNCaP prostate cancer cells. *Clin. Cancer Res.* 2008 Jun 1;14(11):3562-70.

[33] Chajès V, Thiébaut AC, Rotival M, Gauthier E, Maillard V, Boutron-Ruault MC, Joulin V, Lenoir GM, Clavel-Chapelon F. Association between serum trans-monounsaturated fatty acids and breast cancer risk in the E3N-EPIC Study. *Am. J. Epidemiol.* 2008 Jun 1;167(11):1312-20.

[34] Drăgan S, Nicola T, Ilina R, Ursoniu S, Kimar A, Nimade S, Nicola T. Role of multi-component functional foods in the complex treatment of patients with advanced breast cancer. *Rev. Med. Chir .Soc. Med. Nat. Iasi.* 2007 Oct-Dec;111(4):877-84.

[35] Canadanović-Brunet J, Cetković G, Djilas S, Tumbas V, Bogdanović G, Mandić A, Markov S, Cvetković D, Canadanović V. Radical scavenging, antibacterial, and antiproliferative activities of Melissa officinalis L. extracts. *J. Med. Food.* 2008 Mar;11(1):133-43.

[36] Colomer R, Lupu R, Papadimitropoulou A, Vellón L, Vázquez-Martín A, Brunet J, Fernández-Gutiérrez A, Segura-Carretero A, Menéndez JA. Giacomo Castelvetro's salads. Anti-HER2 oncogene nutraceuticals since the 17th century? *Clin. Transl. Oncol.* 2008 Jan;10(1):30-4.

[37] Hallgren O, Aits S, Brest P, Gustafsson L, Mossberg AK, Wullt B, Svanborg C. Apoptosis and tumor cell death in response to HAMLET (human alpha-lactalbumin made lethal to tumor cells). *Adv. Exp. Med. Biol.* 2008;606:217-40.

[38] Dogan S, Hu X, Zhang Y, Maihle NJ, Grande JP, Cleary MP. Effects of high-fat diet and/or body weight on mammary tumor leptin and apoptosis signaling pathways in MMTV-TGF-alpha mice. *Breast Cancer Res.* 2007;9(6):R91.

[39] Ashour HM. Antibacterial, antifungal, and anticancer activities of volatile oils and extracts from stems, leaves, and flowers of Eucalyptus sideroxylon and Eucalyptus torquata. *Cancer Biol. Ther.* 2008 Mar;7(3):399-403.

[40] Wei N, Mi MT, Wang B, Zhu JD, Zhu DP, Yuan JL. Effects of different dietary fatty acid on expression of nuclear receptor genes in breast cancer of rats. Zhonghua Yu Fang Yi Xue Za Zhi. 2007 Jul;41(4):271-6.

[41] Pierce JP, Newman VA, Natarajan L, Flatt SW, Al-Delaimy WK, Caan BJ, Emond JA, Faerber S, Gold EB, Hajek RA, Hollenbach K, Jones LA,

Karanja N, Kealey S, Madlensky L, Marshall J, Ritenbaugh C, Rock CL, Stefanick ML, Thomson C, Wasserman L, Parker BA. Telephone counseling helps maintain long-term adherence to a high-vegetable dietary pattern. *J. Nutr.* 2007 Oct;137(10):2291-6.

[42] Escrich E, Moral R, Grau L, Costa I, Solanas M. Molecular mechanisms of the effects of olive oil and other dietary lipids on cancer. *Mol. Nutr. Food Res.* 2007 Oct;51(10):1279-92.

[43] Allen NE, Grace PB, Ginn A, Travis RC, Roddam AW, Appleby PN, Key T. Phytanic acid: measurement of plasma concentrations by gas-liquid chromatography-mass spectrometry analysis and associations with diet and other plasma fatty acids. *Br. J. Nutr.* 2008 Mar;99(3):653-9.

[44] Chen J, Power KA, Mann J, Cheng A, Thompson LU. Flaxseed alone or in combination with tamoxifen inhibits MCF-7 breast tumor growth in ovariectomized athymic mice with high circulating levels of estrogen. *Exp. Biol. Med.* (Maywood). 2007 Sep;232(8):1071-80.

[45] Turck D. Later effects of breastfeeding practice: the evidence. Nestle Nutr Workshop Ser Pediatr Program. 2007;60:31-9.

[46] Drewnowski A, Henderson SA, Cockroft JE. Genetic sensitivity to 6-n-propylthiouracil has no influence on dietary patterns, body mass indexes, or plasma lipid profiles of women. *J. Am. Diet Assoc.* 2007 Aug;107(8):1340-8.

[47] Midthune D, Kipnis V, Freedman LS, Carroll RJ. Binary regression in truncated samples, with application to comparing dietary instruments in a large prospective study. *Biometrics.* 2008 Mar;64(1):289-98.

[48] Djuric Z, Chen G, Ren J, Venkatramanamoorthy R, Covington CY, Kucuk O, Heilbrun LK. Effects of high fruit-vegetable and/or low-fat intervention on breast nipple aspirate fluid micronutrient levels. *Cancer Epidemiol. Biomarkers Prev.* 2007 Jul;16(7):1393-9.

[49] Nagata C, Iwasa S, Shiraki M, Sahashi Y, Shimizu H. Association of maternal fat and alcohol intake with maternal and umbilical hormone levels and birth weight. *Cancer Sci.* 2007 Jun;98(6):869-73.

[50] Ray A, Nkhata KJ, Grande JP, Cleary MP. Diet-induced obesity and mammary tumor development in relation to estrogen receptor status. *Cancer Lett.* 2007 Aug 18;253(2):291-300.

[51] Grainger EM, Kim HS, Monk JP, Lemeshow SA, Gong M, Bahnson RR, Clinton SK. Consumption of dietary supplements and over-the-counter and prescription medications in men participating in the Prostate Cancer Prevention Trial at an academic center. *Urol. Oncol.* 2008 Mar-Apr;26(2):125-32.

[52] Palmer J, Venkateswaran V, Fleshner NE, Klotz LH, Cox ME. The impact of diet and micronutrient supplements on the expression of neuroendocrine markers in murine Lady transgenic prostate. *Prostate.* 2008 Mar 1;68(4):345-53.

[53] Narita S, Tsuchiya N, Saito M, Inoue T, Kumazawa T, Yuasa T, Nakamura A, Habuchi T. Candidate genes involved in enhanced growth of human prostate cancer under high fat feeding identified by microarray analysis. *Prostate.* 2008 Feb 15;68(3):321-35.

[54] Stacewicz-Sapuntzakis M, Borthakur G, Burns JL, Bowen PE. Correlations of dietary patterns with prostate health. *Mol. Nutr. Food Res.* 2008 Jan;52(1):114-30.

[55] Venkateswaran V, Haddad AQ, Fleshner NE, Fan R, Sugar LM, Nam R, Klotz LH, Pollak M. Association of diet-induced hyperinsulinemia with accelerated growth of prostate cancer (LNCaP) xenografts. *J. Natl. Cancer Inst.* 2007 Dec 5;99(23):1793-800.

[56] Escrich E, Moral R, Grau L, Costa I, Solanas M. Molecular mechanisms of the effects of olive oil and other dietary lipids on cancer. *Mol. Nutr. Food Res.* 2007 Oct;51(10):1279-92.

[57] Allen NE, Grace PB, Ginn A, Travis RC, Roddam AW, Appleby PN, Key T. Phytanic acid: measurement of plasma concentrations by gas-liquid chromatography-mass spectrometry analysis and associations with diet and other plasma fatty acids. *Br. J. Nutr.* 2008 Mar;99(3):653-9.

Chapter 6

LIVER FLUKE AND CHOLANGIOCARCINOMA

INTRODUCTION

Parasitic infestation is a common infectious disease in the tropical countries. There are several kinds of parasitic infestations. Different clinical manifestations can be seen in different kinds of parasitic infestations. Normally, parasite means living things that have mode of living as parasitism. Parasitism means getting usefulness from the others, namely hosts, without contribution of any advantage to the hosts. Sometimes, parasites can also make disorder to the host. Of several kinds of parasite, liver fluke is widely mentioned as parasite that can lead to the cancer. This is of interest because there are only few species that cause cancer.

Liver fluke has its scientific name as *Opisthorchis viverrini*. This parasite is classified as a trematode, a flatworm with mouth but without anus. Its external appearance seems like a leaf. The habitat of this parasite in human beings is in biliary tract, therefore, this parasite is named as liver fluke. The endemic area of this parasite is in the Southeast Asia countries. In Northeastern region of Thailand, the peak prevalence of this parasite can be seen. The parasite is highly contagious. Its small leaflet shape appearance seems characteristic for this parasite. The infective stage of this parasite is the larvae stage or metacercariae, similar to other kinds of trematode. Due to the nature of trematode, it has to pass intermediate host before reaching the final host, human beings. Snails and freshwater fishes are the important intermediate hosts for this parasite. The infectious stage larvae of this parasite lived in the scale of many fresh water fishes. When human beings in the endemic area ingest those infected fresh water fishes, mainly cyprinoid fishes, without cooking, the larvae can infect into the human body. Then it will migrate

Table 1. Reports on contamination of liver fluke in intermediate host

Authors	Details
Nithiuthai et al [1]	Nithiuthai et al reported on a survey of trematode cercariae in *Bithynia goniomphalos* in northeast Thailand [1].
Nithiuthai et al [2]	Nithiuthai et al reported on a survey of metacercariae in cyprinoid fish in Nakhon Ratchasima, northeast Thailand [2].
Sri-Aroon et al [3]	Sri-Aroon et al studied on freshwater mollusks of medical importance in Kalasin Province, northeast Thailand [3].
Phongsasakulchoti et al [4]	Phongsasakulchoti et al reported on emergence of *Opisthorchis viverrini* cercariae from naturally infected *Bithynia siamensis* goniomphalos [4].
Ngern-klun et al [5]	Ngern-klun et al reported on field investigation of Bithynia funiculata, intermediate host of *Opisthorchis viverrini* in northern Thailand [5].
Krailas et al [6]	Krailas et al reported on cercarial infection in Paludomus petrosus, freshwater snail in Pa La-U Waterfall [6].
Sri-Aroon et al [7]	Sri-Aroon et al reported on a malacological survey in Phang-Nga Province, southern Thailand, pre- and post-Indian Ocean tsunami [7].
Ukong et al [8]	Ukong et al reportd on studies on the morphology of cercariae obtained from freshwater snails at Erawan Waterfall, Erawan National Park, Thailand [8].
Harinasuta and Harinasuta [9]	Harinasuta and Harinasuta firstly reported on Opisthorchis viverrini's life cycle, intermediate hosts, transmission to man and geographical distribution in Thailand [9].
Chanawong and Waikagul [10]	Chanawong and Waikagul reported on laboratory studies on host-parasite relationship of Bithynia snails and the liver fluke, *Opisthorchis viverrini* [10].
Adam et al [11]	Adam et al performed studies on lophocercous cercariae from *Bithynia siamensis* goniomphalus [11].
Upatham and Sukhapanth [12]	Upatham and Sukhapanth reported on field studies on the bionomics of Bithynia siamensis siamensis and the transmission of *Opisthorchis viverrini* in Bangna, Bangkok, Thailand [12].
Sri-aroon et al [13]	Sri-aroon et al reportd on freshwater mollusks at designated areas in eleven provinces of Thailand according to the water resource development projects [13].
Chanawong et al [14]	Chanawong et al reported on detection of shared antigens of human liver flukes *Opisthorchis viverrini* and its snail host, *Bithynia* spp [14].
Tesana [15]	Tesana reported on diversity of mollusks in the Lam Ta Khong reservoir, Nakhon Ratchasima, Thailand [15].
Ditrich et al [16]	Ditrich et al reported on larval stages of medically important flukes (Trematoda) from Vientiane province, Laos [16].
Ditrich et al [17]	Ditrich et al reported on occurrence of some medically important flukes in Nam Ngum water reservoir, Laos [17].

Table 2. Reports on clinical manifestation of cholangiocarcinoma

Authors	Details
Wiwanitkit [18]	Wiwanitkit reported on clinical findings among 62 Thais with cholangiocarcinoma [18]. Wiwanitkit noted that the severe jaundice and hyperalkalinephosphatasemia were the most common clinical manifestation [18].
Kuo et al [19]	Kuo et al reported on nucin-producing cholangiocarcinoma focusing mainly on clinical experience of 24 cases [19].
Chantajitr et al [20]	Chantajitr et al reported on combined hepatocellular and cholangiocarcinoma focusing on clinical features and prognostic study in a Thai population [20].
Kim et al [21]	Kim et al reported on factors predicting concurrent cholangiocarcinomas associated with hepatolithiasis [21].

to the habitat at the biliary tract of that human beings. The chronic infestation can be seen and this result into the formation of cancer at that area. The cholangiocarcinoma is the most well-known cancer due to this parasite infestation. The adult parasite can form the eggs and pass them into the stool. When the infected human beings have poor toilet practice, the eggs will be passed into the water, then infected snails and fresh water fishes orderly.

CHOLANGIOCARCINOMA

Cholangiocarcinoma is the cancer of the biliary tract. In Southeast Asia, the main cause of cholangiocarcinoma is the chornic liver fluke infestation. The chronic irritation is the origin of the cancer. The infection of liver fluke causes inflammation of the biliatry tract and liver biliary canal. Cholangitis can be seen and can progress to neoplasia. The role of nitrosamine in co-carcinogenesis of this cancer to liver fluke is widely mentioned. This cancer is usually fatal because the patient usually visited to the physician at the late stage. The major clinuical manifestation of this cancer is the severe jaundice and hyperalkalinephosphatasemia. There is no present curative rate. The treatment is usually surgical removal, which is usually not complete. The death rate is very high

Table 3. Reports on liver fluke infestation and cholangiocarcinoma

Authors	Details
Dechakhamphu et al [22]	Dechakhamphu et al reportrd on high excretion of etheno adducts in liver fluke-infected patients: protection by praziquantel against DNA damage [22].
Zhou et al [23]	Zhou et al reported on risk factors for intrahepatic cholangiocarcinoma by a case-control study in China [23].
Sandhu et al [24]	Sandhu et al reported on epigenetic DNA hypermethylation in cholangiocarcinoma focusing on potential roles in pathogenesis, diagnosis and identification of treatment targets [24].
Tischoff and Tannapfel [25]	Tischoff and Tannapfel reported on hepatocellular carcinoma and cholangiocarcinoma focusing on different prognosis, pathogenesis and therapy [25].
Mairiang et at [26]	Mairiang et at reported on ultrasound screening for *Opisthorchis viverrini*-associated cholangiocarcinomas [26].
Tischoff et al [27]	Tischoff et al reported on the role of epigenetic alterations in cholangiocarcinoma [27].
Kawanishi et al [28]	Kawanishi et al reported on oxidative and nitrative DNA damage in animals and patients with inflammatory diseases in relation to inflammation-related carcinogenesis [28].
Prawan et al [29]	Prawan et al reported on association between genetic polymorphisms of CYP1A2, arylamine N-acetyltransferase 1 and 2 and susceptibility to cholangiocarcinoma [29].
Stauffer et al [30]	Stauffer et al reported on biliary liver flukes in immigrants in the United States [30].
Juntavee [31]	Juntavee reported on expression of sialyl Lewis(a) relates to poor prognosis in cholangiocarcinoma [31].
Chaimuangraj et al [32]	Chaimuangraj et al reported on experimental investigation of opisthorchiasis-associated cholangiocarcinoma induction in the Syrian hamster - pointers for control of the human disease [32].

CHOLANGIOCARCINOMIA AND LIVER FLUKE

As already mentioned, it is confirmed for the correlation between cholangiocarcinoma and liver fluke infestation. There are many reports on this area. Some interesting reports will be hereby presented in Table 3.

OPISTHORCHIS VIVERRINI EXCRETORY PRODUCT AND NITROSAMINE IN FERMENTED LOCAL FOOD: A SYNERGISTIC EFFECT ON CARCINOGENESIS

1. Introduction

Cholangiocarcinoma is an important hepatobiliary malignancy [33 - 34]. This cancer high its peak highest prevalence in Southeast Asia, especially for the area namely Mae Khong Region. An important underlying factor for cholangiocarcinoma in this area is a fluke infestation, liver fluke (Opisthorchis viverrini) infestation. Liver fluke has its high prevalence in the same setting. This parasite is considered as a flatworm which has its life cycle in the intermediate host, fresh water fish [35]. The human beings get infection by ingested the fresh water fishes contaminating with infective stage larvae of O. viverrini. Raw fresh water fish dishes are famous for the local population in Mae Khong region and this can be the source for liver fluke infestation and consequent cholangiocarcinoma [36]. The excretory product (ES) of the parasite is classified as important inducer for cholangiocarcinoma [35]. The change in expression level of signal transduction genes; pkC, pdgfr alpha, jak 1, eps 8, tgf beta 1i4 and h ras relating to the epidermal growth factor (EGF) or transforming growth factor-beta (TGF-beta) is described as possible pathway for the cholangiocarcinoma development [33 - 37].

Another important underlying factor for cholangiocarcinoma is the exposure to nitrosamine. Indeed, nitrosamine is considered as an important carcinogen. The exposure to nitrosamine among the local populations in Mae Khong Region is very high. The reason is the local lifestyle of eating fermented foods, which contain high level of nitrosamine [38]. Of interest, the famous food is fermented fresh water fish, namely "Pla Ra". Therefore, the risk of the local population in Mae Khong area to develop cholangiocarcinoma is very high due to the eating of raw and fermented fresh water fish [36]. In this work, the author uses systomics approach to study the effect of co-exposure to O. viverrini ES and nitrosamine on the carcinogenesis process focusing on the expression on previously mentioned signal transduction genes.

2. Materials and Methods

The author set the mentioned genes, pkC, pdgfr alpha, jak 1, eps 8, tgf beta 1i4 and h ras as the foci for this study. This work makes use of systemic biology technique for creation of systematic network. First, the search for data on expression level of the studied genes due to *O. viverrini* ES and nitrosamine in the literature was searched. Then the finalized network was created identified each substance and co –exposure effect.

3. Results

A. Data on Expression Level of the Studied Genes Due to O. Viverrini Es and Nitrosamine

Only the up-regulated expression of pkC [37, 39 – 43], eps 8 [37] and tgfbeta 1i4 [37] are confirmed for the case of exposure to *O. viverrini* ES.

For data on expression level of the studied genes due to nitrosamine, there are many evidences on pkC [44 - 51], jak 1 [52 - 53] and h-ras [54 - 55].

B. Network Creating

It can be identified that there are three groups of pathways: a) *O.viverrini* ES exposure based pathway (pkC, eps8 and tgfbeta 1i4), b) nitrosamine exposure based pathway (pkC, jak 1 and h-ras) and c) co *O.viverrini* ES and nitrosamine or common exposure based pathway (pkC).

The derived network is presented in Figure 1.

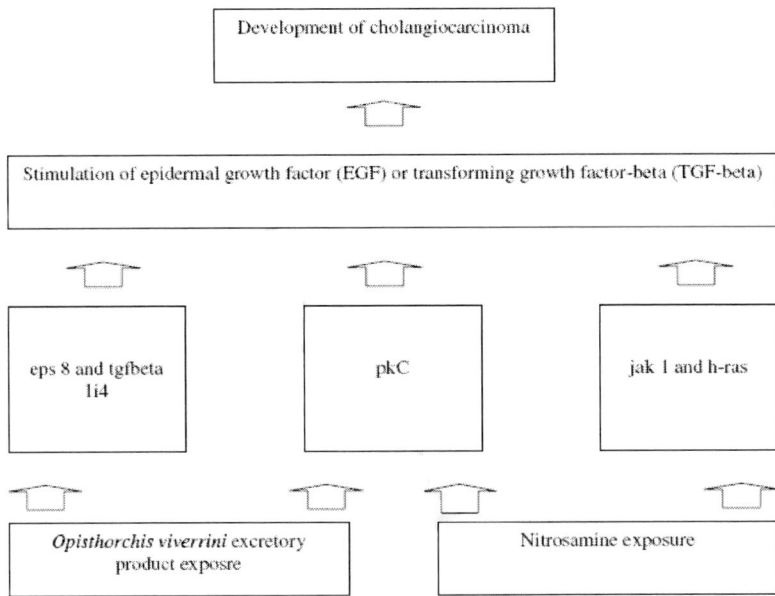

Figure 1. Systomics network for carcinogenesis of cholangiocarcinoma focusing on effect of exposures to O. viverrini ES and nitrosamine.

4. Discussion

An increasing incidence of cholangiocarcinomas has been documented [56]. This increase might be only apparent, due to the progress in investigation and changes in tumor codification [56]. The major clinical sign of cholangiocarcinomas is obstructive jaundice, which is persistent and progressive [56]. In Mae Khong area, the highest prevalence of cholangiocarcinoma is observed and becomes the public threat for the local setting. Of interest, the basic social pathology aspect of this setting has a strong correlation to the very high prevalence. The rooted culture of eating behavior of eating raw and fermented fish is the hallmark for this setting, which is totally difference from the other settings.

It can be seen that there are three separating processes for development of cholangiocarcinoma in this situation. This can confirm the very prevalence of cholangiocarcinoma in the Mae Khong area. It can be seen the alteration of pkC expression is the common pathway indicating that this must be important finding in expressional studies in any settings. In non tropical areas without *O. viverrini*, the aberration on expression of eps8 and tgfbeta 1i4 should not be observed. In those settings, which are usually related to nitrosamine exposure, the aberration of jak 1 and h-ras must be important specific processes.

REFERENCES

[1] Nithiuthai S, Wiwanitkit V, Suwansaksri J, Chaengphukeaw P. A survey of trematode cercariae in Bithynia goniomphalos in northeast Thailand. Southeast *Asian J. Trop. Med. Public Health.* 2002;33 Suppl 3:106-9.

[2] Nithiuthai S, Suwansaksri J, Wiwanitkit V, Chaengphukeaw P. A survey of metacercariae in cyprinoid fish in Nakhon Ratchasima, northeast Thailand. *Southeast Asian J. Trop. Med.Public Health.* 2002;33 Suppl 3:103-5.

[3] Sri-Aroon P, Butraporn P, Limsomboon J, Kerdpuech Y, Kaewpoolsri M, Kiatsiri S. Freshwater mollusks of medical importance in Kalasin Province, northeast Thailand. *Southeast Asian J. Trop. Med. Public Health.* 2005 May;36(3):653-7.

[4] Phongsasakulchoti P, Sri-aroon P, Kerdpuech Y. Emergence of Opisthorchis viverrini cercariae from naturally infected Bithynia (Digoniostoma) siamensis goniomphalos. *Southeast Asian J. Trop. Med. Public Health.* 2005;36 Suppl 4:189-91.

[5] Ngern-klun R, Sukontason KL, Tesana S, Sripakdee D, Irvine KN, Sukontason K. Field investigation of Bithynia funiculata, intermediate host of Opisthorchis viverrini in northern Thailand. *Southeast Asian J. Trop .Med .Public Health.* 2006 Jul;37(4):662-72.

[6] Krailas D, Dechruksa W, Ukong S, Janecharut T. Cercarial infection in Paludomus petrosus, freshwater snail in Pa La-U Waterfall. *Southeast Asian J. Trop. Med. Public Health.* 2003 Jun;34(2):286-90.

[7] Sri-Aroon P, Lohachit C, Harada M, Chusongsang P, Chusongsang Y. Malacological survey in Phang-Nga Province, southern Thailand, pre- and post-Indian Ocean tsunami. *Southeast Asian J. Trop. Med. Public Health.* 2006;37 Suppl 3:104-9.

[8] Ukong S, Krailas D, Dangprasert T, Channgarm P. Studies on the morphology of cercariae obtained from freshwater snails at Erawan Waterfall, Erawan National Park, Thailand. *Southeast Asian J. Trop. Med. Public Health.* 2007 Mar;38(2):302-12.

[9] Harinasuta C, Harinasuta T. Opisthorchis viverrini: life cycle, intermediate hosts, transmission to man and geographical distribution in Thailand. Arzneimittelforschung. 1984;34(9B):1164-7.

[10] Chanawong A, Waikagul J. Laboratory studies on host-parasite relationship of Bithynia snails and the liver fluke, Opisthorchis viverrini. *Southeast Asian J. Trop. Med. Public Health.* 1991 Jun;22(2):235-9.

[11] Adam R, Arnold H, Pipitgool V, Sithithaworn P, Hinz E, Storch V. Studies on lophocercous cercariae from Bithynia siamensis goniomphalus (Prosobranchia: Bithyniidae). *Southeast Asian J. Trop. Med Public Health.* 1993 Dec;24(4):697-700.

[12] Upatham ES, Sukhapanth N. Field studies on the bionomics of Bithynia siamensis siamensis and the transmission of Opisthorchis viverrini in Bangna, Bangkok, Thailand. *Southeast Asian J. Trop. Med. Public Health.* 1980 Sep;11(3):355-8.

[13] Sri-aroon P, Butraporn P, Limsoomboon J, Kaewpoolsri M, Chusongsang Y, Charoenjai P, Chusongsang P, Numnuan S, Kiatsiri S. Freshwater mollusks at designated areas in eleven provinces of Thailand according to the water resource development projects. *Southeast Asian J. Trop. Med Public Health.* 2007 Mar;38(2):294-301.

[14] Chanawong A, Waikagul J, Thammapalerd N. Detection of shared antigens of human liver flukes Opisthorchis viverrini and its snail host, Bithynia spp. *Trop. Med. Parasitol.* 1990 Dec;41(4):419-21.

[15] Tesana S. Diversity of mollusks in the Lam Ta Khong reservoir, Nakhon Ratchasima, Thailand. *Southeast Asian J. Trop .Med. Public Health.* 2002 Dec;33(4):733-8.

[16] Ditrich O, Nasincová V, Scholz T, Giboda M. Larval stages of medically important flukes (Trematoda) from Vientiane province, Laos. Part II. Cercariae. *Ann. Parasitol. Hum. Comp.* 1992;67(3):75-81.

[17] Ditrich O, Scholz T, Giboda M. Occurrence of some medically important flukes (Trematoda: Opisthorchiidae and Heterophyidae) in Nam Ngum water reservoir, Laos. *Southeast Asian J. Trop. Med. Public Health.* 1990 Sep;21(3):482-9.

[18] Wiwanitkit V. Clinical findings among 62 Thais with cholangiocarcinoma. *Trop. Med. Int Health.* 2003 Mar;8(3):228-30.

[19] Kuo CM, Changchien CS, Wu KL, Chuah SK, Chiu KW, Chiu YC, Chou YP, Kuo CH. Mucin-producing cholangiocarcinoma: clinical experience of 24 cases in 16 years *Scand J. Gastroenterol.* 2005 Apr;40(4):455-9.

[20] Chantajitr S, Wilasrusmee C, Lertsitichai P, Phromsopha N. Combined hepatocellular and cholangiocarcinoma: clinical features and prognostic study in a Thai population. *J. Hepatobiliary Pancreat. Surg.* 2006;13(6):537-42.

[21] Kim YT, Byun JS, Kim J, Jang YH, Lee WJ, Ryu JK, Kim SW, Yoon YB, Kim CY. Factors predicting concurrent cholangiocarcinomas associated with hepatolithiasis. *Hepatogastroenterology.* 2003 Jan-Feb;50(49):8-12.
[22] Dechakhamphu S, Yongvanit P, Nair J, Pinlaor S, Sitthithaworn P, Bartsch H. High excretion of etheno adducts in liver fluke-infected patients: protection by praziquantel against DNA damage. *Cancer Epidemiol. Biomarkers Prev.* 2008 Jul;17(7):1658-64.
[23] Zhou YM, Yin ZF, Yang JM, Li B, Shao WY, Xu F, Wang YL, Li DQ. Risk factors for intrahepatic cholangiocarcinoma: a case-control study in China. *World J. Gastroenterol.* 2008 Jan 28;14(4):632-5.
[24] Sandhu DS, Shire AM, Roberts LR. Epigenetic DNA hypermethylation in cholangiocarcinoma: potential roles in pathogenesis, diagnosis and identification of treatment targets. *Liver Int.* 2008 Jan;28(1):12-27.
[25] Tischoff I, Tannapfel A. Hepatocellular carcinoma and cholangiocarcinoma--different prognosis, pathogenesis and therapy. *Zentralbl Chir.* 2007 Aug;132(4):300-5.
[26] Mairiang E, Chaiyakum J, Chamadol N, Laopaiboon V, Srinakarin J, Kunpitaya J, Sriamporn S, Suwanrungruang K, Vatanasapt V. Ultrasound screening for Opisthorchis viverrini-associated cholangiocarcinomas: experience in an endemic area. *Asian Pac. J. Cancer Prev.* 2006 Jul-Sep;7(3):431-3.
[27] Tischoff I, Wittekind C, Tannapfel A. Role of epigenetic alterations in cholangiocarcinoma. *J. Hepatobiliary Pancreat Surg.* 2006;13(4):274-9.
[28] Kawanishi S, Hiraku Y, Pinlaor S, Ma N. Oxidative and nitrative DNA damage in animals and patients with inflammatory diseases in relation to inflammation-related carcinogenesis. *Biol. Chem.* 2006 Apr;387(4):365-72.
[29] Prawan A, Kukongviriyapan V, Tassaneeyakul W, Pairojkul C, Bhudhisawasdi V. Association between genetic polymorphisms of CYP1A2, arylamine N-acetyltransferase 1 and 2 and susceptibility to cholangiocarcinoma. *Eur .J. Cancer Prev.* 2005 Jun;14(3):245-50.
[30] Stauffer WM, Sellman JS, Walker PF. Biliary liver flukes (Opisthorchiasis and Clonorchiasis) in immigrants in the United States: often subtle and diagnosed years after arrival. *J. Travel Med.* 2004 May-Jun;11(3):157-9.
[31] Juntavee A, Sripa B, Pugkhem A, Khuntikeo N, Wongkham S. Expression of sialyl Lewis(a) relates to poor prognosis in cholangiocarcinoma. *World J. Gastroenterol.* 2005 Jan 14;11(2):249-54.
[32] Chaimuangraj S, Thamavit W, Tsuda H, Moore MA. Experimental investigation of opisthorchiasis-associated cholangiocarcinoma induction in

the Syrian hamster - pointers for control of the human disease. *Asian Pac. J. Cancer Prev.* 2003 Apr-Jun;4(2):87-93.

[33] Khan SA, Thomas HC, Davidson BR, Taylor-Robinson SD. Cholangiocarcinoma. *Lancet.* 2005 Oct 8;366(9493):1303-14.

[34] Patel T. Cholangiocarcinoma. Nat Clin Pract Gastroenterol Hepatol. 2006 Jan;3(1):33-42.

[35] Kaewpitoon N, Kaewpitoon SJ, Pengsaa P, Sripa B. Opisthorchis viverrini: the carcinogenic human liver fluke. *World J. Gastroenterol.* 2008 Feb 7;14(5):666-74.

[36] Pairojkul C, Shirai T, Hirohashi S, Thamavit W, Bhudhisawat W, Uttaravicien T, Itoh M, Ito N. Multistage carcinogenesis of liver-fluke-associated cholangiocarcinoma in Thailand. *Princess Takamatsu Symp.* 1991;22:77-86.

[37] Thuwajit P, Chawengrattanachot W, Thuwajit C, Sripa B, May FE, Westley BR, Tepsiri NN, Paupairoj A, Chau-In S. Increased TFF1 trefoil protein expression in Opisthorchis viverrini-associated cholangiocarcinoma is important for invasive promotion. *Hepatol. Res.* 2007 Apr;37(4):295-304.

[38] Mitacek EJ, Brunnemann KD, Suttajit M, Martin N, Limsila T, Ohshima H, Caplan LS. Exposure to N-nitroso compounds in a population of high liver cancer regions in Thailand: volatile nitrosamine (VNA) levels in Thai food. *Food Chem. Toxicol.* 1999 Apr;37(4):297-305.

[39] Wang Y, Roman R, Schlenker T, Hannun YA, Raymond J, Fitz JG. Cytosolic Ca2+ and protein kinase Calpha couple cellular metabolism to membrane K+ permeability in a human biliary cell line. *J. Clin. Invest.* 1997 Jun 15;99(12):2890-7.

[40] Roman RM, Bodily KO, Wang Y, Raymond JR, Fitz JG. Activation of protein kinase Calpha couples cell volume to membrane Cl- permeability in HTC hepatoma and Mz-ChA-1 cholangiocarcinoma cells. *Hepatology.* 1998 Oct;28(4):1073-80.

[41] Kanno N, Glaser S, Chowdhury U, Phinizy JL, Baiocchi L, Francis H, LeSage G, Alpini G. Gastrin inhibits cholangiocarcinoma growth through increased apoptosis by activation of Ca2+-dependent protein kinase C-alpha. *J. Hepatol.* 2001Feb;34(2):284-91.

[42] Gatof D, Kilic G, Fitz JG. Vesicular exocytosis contributes to volume-sensitive ATP release in biliary cells. Am J Physiol Gastrointest *Liver Physiol.* 2004 Apr;286(4):G538-46.

[43] Alpini G, Kanno N, Phinizy JL, Glaser S, Francis H, Taffetani S, LeSage G.Tauroursodeoxycholate inhibits human cholangiocarcinoma growth via

Ca2+-, PKC-, and MAPK-dependent pathways. *Am. J. Physiol. Gastrointest Liver Physiol.* 2004 Jun;286(6):G973-82.
[44] De Minicis S, Candelaresi C, Marzioni M, Saccomano S, Roskams T, Casini A, Risaliti A, Salzano R, Cautero N, di Francesco F, Benedetti A, Svegliati-Baroni G.
[45] Role of endogenous opioids in modulating HSC activity in vitro and liver fibrosis in vivo. *Gut.* 2008 Mar;57(3):352-64.
[46] Z, Xin M, Deng X. Survival function of protein kinase C{iota} as a novel nitrosamine 4-(methylnitrosamino)-1-(3-pyridyl)-1-butanone-activated bad kinase. *J. Biol. Chem.* 2005 Apr 22;280(16):16045-52.
[47] Choudhury S, Krishna M, Bhattacharya RK. Modulation of NDEA activated ras expression and protein kinase C activity by nicotinamide. *Cancer Lett.* 1999 Dec 1;147(1-2):39-44.
[48] Lee YS, Hong SI, Lee MJ, Kim MR, Jang JJ. Differential expression of protein kinase C isoforms in diethylnitrosamine-initiated rat liver. *Cancer Lett.* 1998 Apr 10;126(1):17-22.
[49] Choudhury S, Krishna M, Bhattacharya RK. Activation of ras oncogenes during hepatocarcinogenesis induced by N-nitrosodiethylamine: possible involvement of PKC and GAP. *Cancer Lett.* 1996 Dec 3;109(1-2):149-54.
[50] La Porta CA, Comolli R. Over-expression of protein kinase C delta is associated with a delay in preneoplastic lesion development in diethylnitrosamine-induced rat hepatocarcinogenesis. *Carcinogenesis.* 1995 May;16(5):1233-8.
[51] La Porta CA, Perletti GP, Piccinini F, Comolli R. Analysis of calcium-dependent protein kinase C isoforms in the early stages of diethylnitrosamine-induced rat hepatocarcinogenesis. *Mol. Carcinog.* 1993;8(4):255-63.
[52] Wang YQ, Ikeda K, Ikebe T, Hirakawa K, Sowa M, Nakatani K, Kawada N, Kaneda K. Inhibition of hepatic stellate cell proliferation and activation by the semisynthetic analogue of fumagillin TNP-470 in rats. *Hepatology.* 2000 Nov;32(5):980-9.
[53] Di Sario A, Bendia E, Taffetani S, Marzioni M, Candelaresi C, Pigini P, Schindler U, Kleemann HW, Trozzi L, Macarri G, Benedetti A. Selective Na+/H+ exchange inhibition by cariporide reduces liver fibrosis in the rat. *Hepatology.* 2003 Feb;37(2):256-66.
[54] Zhu B, Liu GT, Wu RS, Strada SJ. Chemoprevention of bicyclol against hepatic preneoplastic lesions. *Cancer Biol .Ther.* 2006 Dec;5(12):1665-73.
[55] Tran H, Yamamoto S, Morimura K, Min W, Mitsuhashi M, Murai T, Mori S, Hosono M, Oohara T, Makino S, Wanibuchi H, Fukushima S. p53 and H-

ras mutations and microsatellite instability in renal pelvic carcinomas of NON / Shi mice treated with N-butyl-N-(4-hydroxybutyl)-nitrosamine: different genetic alteration from urinary bladder carcinoma. *Jpn. J. Cancer Res.* 2001 Dec;92(12):1278-83.

[56] Acalovschi M. Cholangiocarcinoma: risk factors, diagnosis and management.Rom J Intern Med. 2004;42(1):41-58.

Chapter 7

NITROSAMINE

INTRODUCTION [1 – 15]

Nitrite and nitrate salt are widely used as food ingredients in many Asian countries. The purpose of adding these salts into the food is to increase the redness of the meat. This mimicks the non-freshment of that mentioned meat. Using of these salts are acceptable as food additive if used in allowable levels. However, the problem can be seen if used in excessive level. The accepted level for nitrite is below 125 mg/kg of meat and for nitratre is below 500 mg/kg of meat. This is because of the stability of these salts. Indeed, nitrate is more stable than nitrite when considered in redox reaction. This is also related to the theory of carcinogenesis. The accepted level of usage is also relied on this fact. For these two salts, nitrate might relate to more formation of radical.

There are many kinds of foods that these two salts are used. Examples of these foods include dry salted fish (please read more details about this kind of food in another chapter in this book focusing on salt), sausage, fermented fish as well as local pork products. In Southeast Asia, these two salts are widely used and have a famous name as "Din-Pra-Sew". The local problematic pork products that are added with Din-Pra-Sew are the foods namely "Nham", "Moo-Yor" and "Koonchiang." These food are variants of sausage. The problem can be seen if Din-Pra-Sew is used in excessive level, more than acceptable cut-offs. The nitrite and nitrate can make several health effects including headache, fatique, nausea and vomiting.

NITRATE, NITRITE, NITROSAMINE AND CARCINOGENESIS

Considering the reation, both nitrite and nitrate salt can directly interact with amine in the meat and cause the nitrosamine, the problematic substance. Indeed, nitrosamine is well classified as carcinogen. The gastric juice is also a good catalyst for this reaction. This problematic biochemical reaction can be well generated in the stomach. This is also related to the fact that the nitrate and nitrite salts are mentioned for the relationship to the cancer of gastrointestinal tract. Several cancers including liver cancer, stomach cancer as well as esophageal cancer are related to the ingestion of nitrite and nitrate salts. Strong correlation is reported in the literature. In addition, in fermentedfood, the nitrosamine can act accompanied with the contaminated parasite to generate the cancer. This is the specific scenario of liver fluke infestation and cholangiocarcinoma. (Please read more details about this scenario in another chapter in this book.) The complexicity of nitrosamine and liver fluke infestation in the pathogenesis of cholangiocarcinoma is the focused topic in tropical oncology at present. As already mentioned, the free radical generation is belived to be the underlying mechanism for the carcinogenesis. In the present day, there are several attempts to add vitamin C and vitamin E into the food that the nitrite and nitrate salt are used. This is to promote the antioxidant activity [16 – 20]. This is the hope to stop the process of cancer induction [16 – 20]. However, the best way is the avoidance of eating nitrate and nitrited salts added foods.

The interrelationship between nitrate, nitrite, nitrosamine and carcinogenesis is complex and there ae many reports on this topic. Interesting publications are shown in Table 1. Further studied on this area are warranted in tropical medicine.

Table 1. Reports on nitrate, nitrite, nitrosamine in food and carcinogenesis

Authors	Details
Seel et al [21]	Seel et al studied on N-nitroso compounds in two nitrosated food products in southwest Korea [21].
Lintas et al [22]	Lintas et al reported on in vivo stability of nitrite and nitrosamine formation in the dog stomach: effect of nitrite and amine concentration and of ascorbic acid [22].
Mitacek et al [23]	Mitacek et al reported on geographic distribution of liver and stomach cancers in Thailand in relation to estimated dietary intake of nitrate, nitrite, and

	nitrosodimethylamine [23].
Ward et al [24]	Ward et al discussed on processed meat intake, CYP2A6 activity and risk of colorectal adenoma [24].
Sinha et al [25]	Sinha et al reported on development of a food frequency questionnaire module and databases for compounds in cooked and processed meats [25].
McKnight et al [26]	McKnight et al discussed on dietary nitrate in man [26].
Bingham et al [27]	Bingham et al discussed whether increased endogenous formation of N-nitroso compounds in the human colon explain the association between red meat and colon cancer [27].
La Vecchia et al [28]	La Vecchia et al discussed on nitrosamine intake and gastric cancer risk [28].
Mirvish [29]	Mirvish discussed on role of N-nitroso compounds and N-nitrosation in etiology of gastric, esophageal, nasopharyngeal and bladder cancer and contribution to cancer of known exposures to N-nitroso compounds [29].
Jhee and Watanab [30]	Jhee and Watanabe reported on N-nitroso compounds in two nitrosated food products in southwest Korea [30].
Xu et al [31]	Xu et al reported on the relationship between gastric mucosal changes and nitrate intake via drinking water in a high-risk population for gastric cancer in Moping county, China [31].
Packer et al [32]	Packer et al reported on the effect of different sources of nitrate exposure on urinary nitrate recovery in humans and its relevance to the methods of estimating nitrate exposure in epidemiological studies [32].
Palmer and Mathews [33]	Palmer and Mathews discussed on the role of non-nutritive dietary constituents including nitrite and nitrate in carcinogenesis [33].
Weisburger [34]	Weisburger discussed on role of fat, fiber, nitrate, and food additives in carcinogenesis [34].
Mirvish et al [35]	Mirvish et al reportd on disappearance of nitrite from the rat stomach focusing on contribution of emptying and other factors [35].

REFERENCES

[1] Hamon M. Can nitrates lead to indirect toxicity? *Ann. Pharm. Fr.* 2007 Sep;65(5):347-55.
[2] Scotter MJ, Castle L. Chemical interactions between additives in foodstuffs: a review. *Food Addit Contam.* 2004 Feb;21(2):93-124.
[3] Boyce MC. Determination of additives in food by capillary electrophoresis. *Electrophoresis.* 2001 May;22(8):1447-59.
[4] Grudziński IP. Effect of nitrates and nitrites on small intestine. Rocz Panstw Zakl Hig. 1998;49(3):321-30.
[5] Di Matteo V, Esposito E. Methods for the determination of nitrite by high-performance liquid chromatography with electrochemical detection. *J. Chromatogr. A.* 1997 Nov 21;789(1-2):213-9.
[6] Unden G, Becker S, Bongaerts J, Schirawski J, Six S. Oxygen regulated gene expression in facultatively anaerobic bacteria. Antonie Van Leeuwenhoek. 1994;66(1-3):3-22. Hernández E, Huerta T. Evolution of the microbiological parameters of cured ham.
[7] Microbiologia. 1993 Feb;9 Spec No:10-9. Massey RC, Lees D. Surveillance of preservatives and their interactions in foodstuffs. *Food Addit Contam.* 1992 Sep-Oct;9(5):435-40.
[8] Skovgaard N. Microbiological aspects and technological need: technological needs for nitrates and nitrites. *Food Addit Contam.* 1992 Sep-Oct;9(5):391-7.
[9] Lauer K. The history of nitrite in human nutrition: a contribution from German cookery books. *J. Clin. Epidemiol.* 1991;44(3):261-4.
[10] Walker R. Nitrates, nitrites and N-nitrosocompounds: a review of the occurrence in food and diet and the toxicological implications. *Food Addit Contam.* 1990 Nov-Dec;7(6):717-68.
[11] Palmer S, Mathews RA. The role of non-nutritive dietary constituents in carcinogenesis. *Surg. Clin. North Am.* 1986 Oct;66(5):891-915.
[12] Weisburger JH. Role of fat, fiber, nitrate, and food additives in carcinogenesis: a critical evaluation and recommendations. *Nutr. Cancer.* 1986;8(1):47-62.
[13] Hartman PE. Review: putative mutagens and carcinogens in foods. I. Nitrate/nitrite ingestion and gastric cancer mortality. *Environ. Mutagen.* 1983;5(1):111-21.
[14] Issenberg P. Nitrite, nitrosamines, and cancer. Fed Proc. 1976 May 1;35(6):1322-6. Ostergaard K. Nitrate--nitrite--nitrosamine--cancer? Ugeskr Laeger. 1975 Dec 29;138(1):18-20.

[15] Emerick RJ. Consequences of high nitrate levels in feed and water supplies. *Fed Proc.* 1974 May;33(5):1183-7.

[16] Rasheed MH, Beevi SS, Geetha A. Enhanced lipid peroxidation and nitric oxide products with deranged antioxidant status in patients with head and neck squamous cell carcinoma. *Oral Oncol.* 2007 Apr;43(4):333-8.

[17] Chen H, Ward MH, Tucker KL, Graubard BI, McComb RD, Potischman NA, Weisenburger DD, Heineman E. Diet and risk of adult glioma in eastern Nebraska, United States. *Cancer Causes Control.* 2002 Sep;13(7):647-55.

[18] Bunin GR, Kuijten RR, Boesel CP, Buckley JD, Meadows AT. Maternal diet and risk of astrocytic glioma in children: a report from the Childrens Cancer Group (United States and Canada) *Cancer Causes Control.* 1994 Mar;5(2):177-87.

[19] Choi NW, Miller AB, Fodor JG, Jain M, Howe GR, Risch HA, Ruder AM. Consumption of precursors of N-nitroso compounds and human gastric cancer. *IARC Sci Publ.* 1987;(84):492-6.

[20] Weisburger JH, Horn CL. Human and laboratory studies on the causes and prevention of gastrointestinal cancer. Scand. *J.Gastroenterol. Suppl.* 1984;104:15-26.

[21] Seel DJ, Kawabata T, Nakamura M, Ishibashi T, Hamano M, Mashimo M, Shin SH, Sakamoto K, Jhee EC, Watanabe S. N-nitroso compounds in two nitrosated food products in southwest Korea. *Food Chem. Toxicol.* 1994 Dec;32(12):1117-23.

[22] Lintas CL, Clark A, Fox J, Tannenbaum SR, Newberne PM. In vivo stability of nitrite and nitrosamine formation in the dog stomach: effect of nitrite and amine concentration and of ascorbic acid. *Carcinogenesis.* 1982;3(2):161-5.

[23] Mitacek EJ, Brunnemann KD, Suttajit M, Caplan LS, Gagna CE, Bhothisuwan K, Siriamornpun S, Hummel CF, Ohshima H, Roy R, Martin N. Geographic distribution of liver and stomach cancers in Thailand in relation to estimated dietary intake of nitrate, nitrite, and nitrosodimethylamine. *Nutr Cancer.* 2008 Mar-Apr;60(2):196-203.

[24] Ward MH, Cross AJ, Divan H, Kulldorff M, Nowell-Kadlubar S, Kadlubar FF, Sinha R. Processed meat intake, CYP2A6 activity and risk of colorectal adenoma. *Carcinogenesis.* 2007 Jun;28(6):1210-6.

[25] Sinha R, Cross A, Curtin J, Zimmerman T, McNutt S, Risch A, Holden J. Development of a food frequency questionnaire module and databases for compounds in cooked and processed meats. *Mol. Nutr. Food Res.* 2005 Jul;49(7):648-55.

[26] McKnight GM, Duncan CW, Leifert C, Golden MH. Dietary nitrate in man: friend or foe? *Br. J. Nutr.* 1999 May;81(5):349-58.

[27] Bingham SA, Pignatelli B, Pollock JR, Ellul A, Malaveille C, Gross G, Runswick S, Cummings JH, O'Neill IK. Does increased endogenous formation of N-nitroso compounds in the human colon explain the association between red meat and colon cancer? *Carcinogenesis.* 1996 Mar;17(3):515-23.

[28] La Vecchia C, D'Avanzo B, Airoldi L, Braga C, Decarli A. Nitrosamine intake and gastric cancer risk. *Eur. J. Cancer Prev.* 1995 Dec;4(6):469-74.

[29] Mirvish SS. Role of N-nitroso compounds (NOC) and N-nitrosation in etiology of gastric, esophageal, nasopharyngeal and bladder cancer and contribution to cancer of known exposures to NOC. Cancer Lett. 1995 Jun 29;93(1):17-48. Review. Erratum in: *Cancer Lett.* 1995 Nov 6;97(2):271.

[30] Xu G, Song P, Reed PI. The relationship between gastric mucosal changes and nitrate intake via drinking water in a high-risk population for gastric cancer in Moping county, China. *Eur. J. Cancer Prev.* 1992 Oct;1(6):437-43.

[31] Packer PJ, Leach SA, Duncan SN, Thompson MH, Hill MJ. The effect of different sources of nitrate exposure on urinary nitrate recovery in humans and its relevance to the methods of estimating nitrate exposure in epidemiological studies. *Carcinogenesis.* 1989 Nov;10(11):1989-96.

[32] Palmer S, Mathews RA. The role of non-nutritive dietary constituents in carcinogenesis. *Surg. Clin. North Am.* 1986 Oct;66(5):891-915.

[33] Weisburger JH. Role of fat, fiber, nitrate, and food additives in carcinogenesis: a critical evaluation and recommendations. *Nutr. Cancer.* 1986;8(1):47-62.

[34] Lintas CL, Clark A, Fox J, Tannenbaum SR, Newberne PM. In vivo stability of nitrite and nitrosamine formation in the dog stomach: effect of nitrite and amine concentration and of ascorbic acid. *Carcinogenesis.* 1982;3(2):161-5.

[35] Mirvish SS, Patil K, Ghadirian P, Kommineni VR. Disappearance of nitrite from the rat stomach: contribution of emptying and other factors. *J. Natl. Cancer Inst.* 1975 Apr;54(4):869-75.

Chapter 8

SALTED AND FERMENTED FOOD

INTRODUCTION

Salt is a basic ingredient for food. Salt has been produced and used in kitchen science for a very long time. The inquiry of salt can be from sea or dirt. There are two main kinds of salt, marine salt and dirt salt. Marine salt derives from the sea water exposed to sunlight. This has the iodine content as well. This is useful to prevent the condtion namely endemic goiter. While dirt salt derived from underground water. It has no iodine. Therefore, supplementation of iodine in dirt salt is recommended. The dirt salt is produced in land lock country. Of inerest, this area exposes to the prolem of iodine deficiency. The good example is the scenario in the Northeastern Region of Thailand. At present, classical production of salt in salt field can be seen in seashore cities in developing countries. This process is not existed in developed countries. Of interest, salt is the cystal of sodium chloride (NaCL). Therefore, it gives two main elements, sodium and choride to the consumer. These two elements are the important electrolytes for human beings. The water regulation in human bodies remains relating to sodium. However, excessive ingestion of salt can be problematic. This can bring several health disoders. The well-known ones include hypertension. The hypernatremia is the condition describing for the toxicity due to excessive sodium in human body. In addition, salt is also mentioned as an important substance contributing to the carcinogenesis. The gastric cancer is mentioned for the relationship to excessive ingestion of salt.

SALT AND CARCINOGENESIS

Of interest, the areas that local population eats salted food such as dry salted fishes have high incidence of gastric cancer. In the countries that the locol people love to eat dry saled vegetable, the incidence of nasopharyngeal cancer is very high. However, this might be due to the high incidence of EBV virus infection in that area. This must be concerned as a confounding factor. Also, high prevalence of other cancers such as gastric cancer and esophageal cancer can be detected in this area. The presence of the nitrosamine in salted fish might be a clue for this observation. (More information on nitrosamine can be read in another specific chapter in this book.) In addition to dry sated fish, salted vegetable is also widely ingested. Ingestion of salted vegetable accompanied with tobacco smoking and alcoholic beverage drinking has strong relationship to gastric cancer, nasopharyngeal cancer and esophageal cancer. The nitrosamine can also be detected in salted vegetable.Kimchi is the most well-known food in this category. This is owing to the adding of specific amine substance into the salted vegetable in the production process of salted vegetable to increase appetite. Important reports on salt and carcinogenesis are presented in Table 1.

Table 1. Reports on salt and carcinogenesis

Authors	Details
Rogers et al [1]	Rogers et al mentioned that *Helicobacter pylori* but not high salt induces gastric intraepithelial neoplasia in B6129 mice [1].
Nan et al [2]	Nan et al reported that Kimchi and soybean pastes were risk factors of gastric cancer [2].
De Stefani et al [3]	De Stefani et al discussed on dietary patterns and risk of gastric cancer owing to their case-control study in Uruguay [3].
Nakaji et al [4]	Nakaji et al reported on the relationship between mineral and trace element concentrations in drinking water and gastric cancer mortality in Japan [4].
Wada et al [5]	Wada et al reported on effects of catechol, sodium chloride and ethanol either alone or in combination on gastric carcinogenesis in rats pretreated with N-methyl-N'-nitro-N-nitrosoguanidine [5].
Mori et al [6]	Mori et al noted for lack of promotion of urinary bladder carcinogenesis by sodium bicarbonate and/or

Authors	Details
Nishikawa et al [7]	L-ascorbic acid in male ODS/Shi-od/od rats synthesizing alpha 2 mu-globulin but not L-ascorbic acid [6]. Nishikawa et al reported on dose-dependent promotion effects of potassium chloride on glandular stomach carcinogenesis in rats after initiation with N-methyl-N'-nitro-N-nitrosoguanidine and the synergistic influence with sodium chloride [7].
Nishikawa et al [8]	Nishikawa et al reported on effects of caffeine on glandular stomach carcinogenesis induced in rats by N-methyl-N'-nitro-N-nitrosoguanidine and sodium chloride [8].
Nishikawa et al [9]	Nishikawa et al reported on effects of hickory smoke condensate on gastric carcinogenesis in Wistar rats after treatment with N-methyl-N'-nitro-N-nitrosoguanidine and sodium chloride [9].
Correa [10]	Correa discussed on human gastric carcinogenesis expressing a multistep and multifactorial process [10].
Lyon and Mahoney [11]	Lyon and Mahoney discussed on fried foods and the risk of colon cancer [11].
van Berge Henegouwen et al [12]	van Berge Henegouwen et al reported on effect of long term lactulose ingestion on secondary bile salt metabolism in man: potential protective effect of lactulose in colonic carcinogenesis [12].
Takahashi et al [13]	Takahashi et al reported on effect of high salt diet on rat gastric carcinogenesis induced by N-methyl-N'-nitro-N-nitrosoguanidine [13].

REFERENCES

[1] Rogers AB, Taylor NS, Whary MT, Stefanich ED, Wang TC, Fox JG. Helicobacter pylori but not high salt induces gastric intraepithelial neoplasia in B6129 mice. *Cancer Res.* 2005 Dec 1;65(23):10709-15.

[2] Nan HM, Park JW, Song YJ, Yun HY, Park JS, Hyun T, Youn SJ, Kim YD, Kang JW, Kim H. Kimchi and soybean pastes are risk factors of gastric cancer. *World J. Gastroenterol.* 2005 Jun 7;11(21):3175-81.

[3] De Stefani E, Correa P, Boffetta P, Deneo-Pellegrini H, Ronco AL, Mendilaharsu M. Dietary patterns and risk of gastric cancer: a case-control study in Uruguay. *Gastric Cancer.* 2004;7(4):211-20.

[4] Nakaji S, Fukuda S, Sakamoto J, Sugawara K, Shimoyama T, Umeda T, Baxter D. Relationship between mineral and trace element concentrations in drinking water and gastric cancer mortality in Japan. *Nutr. Cancer.* 2001;40(2):99-102.

[5] Wada S, Hirose M, Shichino Y, Ozaki K, Hoshiya T, Kato K, Shirai T. Effects of catechol, sodium chloride and ethanol either alone or in combination on gastric carcinogenesis in rats pretreated with N-methyl-N'-nitro-N-nitrosoguanidine. *Cancer Lett.* 1998 Jan 30;123(2):127-34.

[6] Mori S, Murai T, Hosono M, Machino S, Makino S, Chou C, Fukushima S. Lack of promotion of urinary bladder carcinogenesis by sodium bicarbonate and/or L-ascorbic acid in male ODS/Shi-od/od rats synthesizing alpha 2 mu-globulin but not L-ascorbic acid. *Food Chem. Toxicol.* 1997 Aug;35(8):783-7.

[7] Nishikawa A, Furukawa F, Mitsui M, Enami T, Imazawa T, Ikezaki S, Takahashi M. Dose-dependent promotion effects of potassium chloride on glandular stomach carcinogenesis in rats after initiation with N-methyl-N'-nitro-N-nitrosoguanidine and the synergistic influence with sodium chloride. *Cancer Res.* 1995 Nov 15;55(22):5238-41.

[8] Nishikawa A, Furukawa F, Imazawa T, Ikezaki S, Hasegawa T, Takahashi M. Effects of caffeine on glandular stomach carcinogenesis induced in rats by N-methyl-N'-nitro-N-nitrosoguanidine and sodium chloride. *Food Chem. Toxicol.* 1995 Jan;33(1):21-6.

[9] Nishikawa A, Furukawa F, Imazawa T, Toyoda K, Mitsui M, Hasegawa T, Takahashi M. Effects of hickory smoke condensate on gastric carcinogenesis in Wistar rats after treatment with N-methyl-N'-nitro-N-nitrosoguanidine and sodium chloride. *Food Chem. Toxicol.* 1993 Jan;31(1):25-30.

[10] Correa P. Human gastric carcinogenesis: a multistep and multifactorial process--First American Cancer Society Award Lecture on Cancer Epidemiology and Prevention. *Cancer Res.* 1992 Dec 15;52(24):6735-40.

[11] Lyon JL, Mahoney AW. Fried foods and the risk of colon cancer. Am J Epidemiol. 1988 Nov;128(5):1000-6.

[12] van Berge Henegouwen GP, van der Werf SD, Ruben AT. Effect of long term lactulose ingestion on secondary bile salt metabolism in man: potential protective effect of lactulose in colonic carcinogenesis. *Gut.* 1987 Jun;28(6):675-80.

[13] Takahashi M, Kokubo T, Furukawa F, Kurokawa Y, Tatematsu M, Hayashi Y. Effect of high salt diet on rat gastric carcinogenesis induced by N-methyl-N'-nitro-N-nitrosoguanidine. *Gann.* 1983 Feb;74(1):28-34.

Chapter 9

FOOD ADDITIVE: DYE AND FAVOR

INTRODUCTION [1 – 10]

Food additive is a widely used substance at present. In kitchen science, adding of addititive is to increase favour, smell and taste.

Several kinds of additives are produced. Marketing of these additives are high and use many procedures, which are usually overwhelming on usefulness and safety
of these additives. Standard additives are acceptable for using in kitchen sicence. No serious adverse effect on health for consumers is mentioned. However, the problem is usually owing to non standard additives. These can cause several health effects on the consumers. In this chapter, the author will discuss on two common problematic additives: dye and saccharin. These two mentioned substances are widely used in many developing countries and illegally applied for sometimes. The problems emerge from these two substances are of interest and concern for present medicine. Also, there are other kinds of problematic non additive but contaminated substances due to production process such as antibiotic and beta-blocker. These issues are the present focus in food safety.

DYE

Dye is a substance that can make color. It is usually a chemical. Basically, dye is used for modifying the color of the food. This is concordant with the basic concept that bright color food usually attracts to the consumers. Generally, standard dye for food is acceptable as additive. Several colors can be gnerated by

Table 1. Reports on dye contamination in food and carcinogenesis

Authors	Details
Gupta et al [11]	Gupta et al reported on tumor promotion by metanil yellow and malachite green during rat hepatocarcinogenesis is associated with dysregulated expression of cell cycle regulatory proteins [11].
Sundarrajan et al [12]	Sundarrajan et al reported on overexpression of G1/S cyclins and PCNA and their relationship to tyrosine phosphorylation and dephosphorylation during tumor promotion by metanil yellow and malachite green [12].
Rao and Fernandes [13]	Rao and Fernandes reported on progressive effects of malachite green at varying concentrations on the development of N-nitrosodiethylamine induced hepatic preneoplastic lesions in rats [13].
Fernandes and Rao [14]	Fernandes and Rao discussed on dose related promoter effect of metanil yellow on the development of hepatic pre-neoplastic lesions induced by N-nitrosodiethylamine in rats [14].
Westmoreland and Gatehouse [15]	Westmoreland and Gatehouse reported on the differential clastogenicity of Solvent Yellow 14 and FD and C Yellow No. 6 in vivo in the rodent micronucleus test [15].
Najem et al [16]	Najem et al discussed on life time occupation, smoking, caffeine, saccharin, hair dyes and bladder carcinogenesis [16].

standard dyes. However, the applying of non standard dyes as food additives is problematic in public health. In many developing countries, illegal non standard dsyes are used as food additives. These can lead several health problems to the consumers. These dyes can cause diseases. In general, dye is mainly added into candy, cracker and softdrink. This is usually relating to the attractive purpose to the pediatric group of consumers. Indeed, this kind of foods is the main food for children and children are the targeted population for marketing. Children love candy, crakcer and softdrink. They usually love those ones with bright color. This food might be not widely ingested by the adults. The children usually get toxic when ingest foods with non standard dye. Focusing on the adult population, they usually buy inexpensive foods. These mentioned foods can be inexpensive but contaminated. Therefore, the adult usually get toxic due to this fact. Theoretically, standard dye can be formed natural or synthetic procedures. The source of natural dyes is usually from plant and animal. Several plants can give dyes. These natural dyes are considered safe and should be used. It is presently promoted to use plant derived dyes as food additives. For synthetic dyes, it is acceptable if used in limited acceptable level. However, the problem can be detected if excessive use is performed. By the way, illegal non standard synthetic dye is usually problematic and more widely used. This is due to the fact that these problematic dyes are not expensive therefore sticky food producers use these illegal dyes as food additives. These dyes can result in cancer. The urinary bladder cancer is the widely quoted cancer in this case. There are many reports on this topic. Inportant publications will be demostrated in Table 1. As a conclusion, natural dye is better than synthetic dye. However, the consumers are usually not the ones who prepare the foods. Therefore, it is suggested that consumers should select non dyed, not bright colored, food or food with less bright color.

SACCHARIN

Saccharin is a substance that can give sweet taste similar to sugar. However, saccharin provides no energy but toxicity. Sugar is a carbohydrate that acts as a primary source of energy for human beings. But saccharin is not. It give on toxic sweet taste. It should be noted that saccharin gives more sweet taste than sugar when compared at the same quantity. This can be said that sweet but toxic. In addition, saccharin is cheaper. Therefore, sticky food producers add saccharin instead of sugar into their foods. There are some reports on carcinogenesis of saccharin in animal model studies. Many animal cancers are related to exposre to saccharin. However, in human beings, the evidence is not strong. Saccharin and bladder cancer is widely proposed. This is believed to be due to excretion of saccharin via kidney and further accumulation in the urinary bladder. However, as already mentioned, the interrelationship is not well demonstrated. Important publications on saccharin and carcinogenesis will be discussed in Table 2. Further studies on this area are warranted in present toxicology research. To verify the safety and carcinogenicity of saccharin is useful for further making decision to use this sugar – free sweetener as an alternative to diabetic patients.

Table 2. Reports on saccharin in food and carcinogenesis

Authors	Details
Parfett [17]	Parfett reported on combined effects of tumor promoters and serum on proliferin mRNA induction: a biomarker sensitive to saccharin, 2,3,7,8-TCDD, and other compounds at minimal concentrations promoting C3H/10T1/2 cell transformation [17].
Bell et al [18]	Bell et al reported on carcinogenicity of saccharin in laboratory animals and humans [18].
Weihrauch et al [19]	Weihrauch et al discussed whether artificial sweeteners potentially carcinogenic[19].
Turner et al [20]	Turner et al reported that the male rat carcinogens limonene and sodium saccharin wer not mutagenic to male Big Blue rats [20]
Irie et al [21]	Irie et al reported on classical conditioning of oxidative DNA damage in rats including the saccharin exposure [21].

REFERENCES

[1] Takayama S, Thorgeirsson UP, Adamson RH. Chemical carcinogenesis studies in nonhuman primates. *Proc. Jpn. Acad Ser. B Phys. Biol. Sci.* 2008;84(6):176-88.

[2] Thorgeirsson UP, Dalgard DW, Reeves J, Adamson RH. Tumor incidence in a chemical carcinogenesis study of nonhuman primates. *Regul. Toxicol. Pharmacol.* 1994 Apr;19(2):130-51.

[3] Schoeffner DJ, Thorgeirsson UP. Susceptibility of nonhuman primates to carcinogens of human relevance. *In Vivo.* 2000 Jan-Feb;14(1):149-56.

[4] Thorgeirsson SS, Davis CD, Schut HA, Adamson RH, Snyderwine EG. Possible relationship between tissue distribution of DNA adducts and genotoxicity of food-derived heterocyclic amines. *Princess Takamatsu Symp.* 1995;23:85-92.

[5] National Toxicology Program. NTP Toxicology and Carcinogenesis Studies of Coumarin (CAS No. 91-64-5) in F344/N Rats and B6C3F1 Mice (Gavage Studies). *Natl. Toxicol. Program. Tech .Rep. Ser.* 1993 Sep;422:1-340.

[6] Dalgard DW, Thorgeirsson UP, Adamson RH. Laparoscopy as a means of monitoring liver tumor induction in nonhuman primates. *Princess Takamatsu. Symp.* 1995;23:268-73.

[7] National Toxicology Program. NTP Toxicology and Carcinogenesis Studies of Phenolphthalein (CAS No. 77-09-8) in F344/N Rats and B6C3F1 Mice (Feed Studies). *Natl. Toxicol. Program Tech. Rep. Ser.* 1996 Nov;465:1-354.

[8] National Toxicology Program. Toxicology and Carcinogenesis Studies of 5,5-Diphenylhydantoin (CAS No. 57-41-0) (Phenytoin) in F344/N Rats and B6C3F1 Mice (Feed Studies). *Natl. Toxicol. Program. Tech. Rep. Ser.* 1993 Nov;404:1-303.

[9] National Toxicology Program. NTP Toxicology and Carcinogenesis Studies of C.I. Direct Blue 218 (CAS No. 28407-37-6) in F344/N Rats and B6C3F1 Mice (Feed Studies). *Natl. Toxicol. Program .Tech .Rep. Ser.* 1994 Feb;430:1-280.

[10] National Toxicology Program. NTP Toxicology and Carcinogenesis Studies of Nickel Oxide (CAS No. 1313-99-1) in F344 Rats and B6C3F1 Mice (Inhalation Studies). *Natl. Toxicol. Program. Tech. Rep. Ser.* 1996 Jul;451:1-381.

[11] Gupta S, Sundarrajan M, Rao KV. Tumor promotion by metanil yellow and malachite green during rat hepatocarcinogenesis is associated with dysregulated expression of cell cycle regulatory proteins. *Teratog. Carcinog. Mutagen.* 2003;Suppl 1:301-12.

[12] Sundarrajan M, Fernandis AZ, Subrahmanyam G, Prabhudesai S, Krishnamurthy SC, Rao KV. Overexpression of G1/S cyclins and PCNA and their relationship to tyrosine phosphorylation and dephosphorylation during tumor promotion by metanil yellow and malachite green. *Toxicol. Lett.* 2000 Jul 27;116(1-2):119-30.

[13] Rao KV, Fernandes CL. Progressive effects of malachite green at varying concentrations on the development of N-nitrosodiethylamine induced hepatic preneoplastic lesions in rats. *Tumori.* 1996 May-Jun;82(3):280-6.

[14] Fernandes C, Rao KV. Dose related promoter effect of metanil yellow on the development of hepatic pre-neoplastic lesions induced by N-nitrosodiethylamine in rats. *Indian J. Med. Res.* 1994 Sep;100:140-9.

[15] Westmoreland C, Gatehouse DG. The differential clastogenicity of Solvent Yellow 14 and FD and C Yellow No. 6 in vivo in the rodent micronucleus test (observations on species and tissue specificity). *Carcinogenesis.* 1991 Aug;12(8):1403-7.

[16] Najem GR, Louria DB, Seebode JJ, Thind IS, Prusakowski JM, Ambrose RB, Fernicola AR. Life time occupation, smoking, caffeine, saccharin, hair dyes and bladder carcinogenesis. *Int. J. Epidemiol.* 1982 Sep;11(3):212-7.

[17] Parfett CL. Combined effects of tumor promoters and serum on proliferin mRNA induction: a biomarker sensitive to saccharin, 2,3,7,8-TCDD, and other compounds at minimal concentrations promoting C3H/10T1/2 cell transformation. *J. Toxicol. Environ. Health* A. 2003 Oct 24;66(20):1943-66.

[18] Bell W, Clapp R, Davis D, Epstein S, Farber E, Fox DA, Holub B, Jacobson MF, Lijinsky W, Millstone E, Reuber MD, Suzuki D, Temple NJ. Carcinogenicity of saccharin in laboratory animals and humans: letter to Dr. Harry Conacher of Health Canada. *Int. J. Occup. Environ. Health.* 2002 Oct-Dec;8(4):387-93.

[19] Weihrauch MR, Diehl V, Bohlen H. Artificial sweeteners--are they potentially carcinogenic? Med Klin (Munich). 2001 Nov 15;96(11):670-5.

[20] Turner SD, Tinwell H, Piegorsch W, Schmezer P, Ashby J. The male rat carcinogens limonene and sodium saccharin are not mutagenic to male Big Blue rats.*Mutagenesis.* 2001 Jul;16(4):329-32.

[21] Irie M, Asami S, Nagata S, Miyata M, Kasai H. Classical conditioning of oxidative DNA damage in rats. *Neurosci. Lett.* 2000 Jul 7;288(1):13-6.

Chapter 10

ENVIRONMENTAL CONTAMINANTS

INTRODUCTION

In the present day, due to the urbanization, pollution can be seen in any cities. Pollutantion can be either water, air or soil pollutions. Relationship between pollutant and food can be proposed. Because food comes from farm animals or plants which live in the environment, these can be contaminated by pollutants if there are any pollution in that area. This is the new public health focus in the present day. This problem can be seen in many developed countries. Developing countries which pass the phase of new industrilization can also face up this problem. It is the global concern at present.

Pollutants can be contaminated into the food in several methods. The first methoid is the direct contamination. This means the pollutant in the environment, water or air or soil, directly contact to the food and contaminte into that food. As a result, food contamination can be expected. In addition, the problematic substance can be contaminted into the food in many other conditions. In industrial process, the direct contamination of toxic substances can be expected. The 3-monochloropropane-1,2-diol (MCPD) contamination, which is related to cancer of breast and kindey are reported. The contamination of dioxin is another famous scenario. Dioxin can be the underlying factor for developing of respiratory tract cancer. Contamination of substances due to agricultural processes can also be seen. The usage of pesticide and insecticide can make the contamination into the food. These can be problematic. This situation is related to many cancers including lymphoma, leukemia, stomach cancer, skin cancer and prostate cancer. The second method of contamination is callded indirect contamination. This is usually owing to the indirect passing into the food chain. Generally, the food chain is the basic concept in nutritional medicine. Focusing on a food chain, the larger animal eats smaller ones. Once, larger animal eats the smaller ones, they get nutrients as well as toxic substances accumulated in thoese smaller animals. The accumation of the toxic substances in the larger animal can be expected and can be increased in level when time passes. In the food chain, human beings are usually the terminal of the tract. Human beings eat several animals, therefore, they can accumulate several toxic substances in their bodies. In this chaper, the author will discuss on some important problematic substances including MCPD, dioxin, pesticide and insecticide, which are all mentioned for the relationship to carcinogenicity.

MCPD

The MCPD contamination is mentioned for the strong correlation to cancer of breast and kidney. The important reports on this area are listed in Table 1.

Table 1. Reports on MCPD and carcinogenesis.

Authors	Details
Cho et al [1]	Cho et al reported on carcinogenicity study of 3-monochloropropane-1,2-diol in Sprague-Dawley rats [1].
Cho et al [2]	Cho et al reported on subchronic toxicity study of 3-monochloropropane-1,2-diol administered by drinking water to B6C3F1 mice [2].

DIOXIN

Dioxin contamination became an important public health interest in the past decade. The carcinogenesis of this substance is confirmed in many reports. Some important reports are hereby quoted in Table 2.

Table 2. Reports on dioxin and carcinogenesis

Authors	Details
Kuhn et al [3]	Kuhn et al reported on dDetermination of polycyclic aromatic hydrocarbons in smoked pork by effect-directed bioassay with confirmation by chemical analysis [3].
Bellocq et al [4]	Bellocq et al reported on high potency of bioactivation of 2-amino-1-methyl-6-phenylimidazo[4,5-b]pyridine (PhIP) in mouse colon epithelial cells with Apc(Min) mutation [4].
Ye and Leung [5]	Ye and Leung reported on effect of dioxin exposure on aromatase expression in ovariectomized rats [5].
Belpomme et al [6]	Belpomme et al reported on the multitude and diversity of environmental carcinogens including dioxin [6].
Oesch-Bartlomowicz et al [7]	Oesch-Bartlomowicz et al reported on aryl hydrocarbon receptor activation by cAMP versus dioxin [7].
Hardell et al [8]	Hardell et al reported on increased concentrations of octachlorodibenzo-p-dioxin in cases with breast cancer [8].

PESTICIDE AND INSECTICIDE

As already mentioned, application of pesticide and insecticide can lead to the contamination into the food. These can be problematic. This scenario is related to many cancers including lymphoma, leukemia, stomach cancer, skin cancer and prostate cancer. Important reports on this topic are summarized in Table 3.

Table 3. Reports on pesticide, insecticide and carcinogenesis

Authors	Details
Wong et al [9]	Wong et al reported on polymorphisms in metabolic GSTP1 and DNA-repair XRCC1 genes with an increased risk of DNA damage in pesticide-exposed fruit growers [9].
Dorsey et al [10]	Dorsey et al reportd on mitogenic and cytotoxic effects

Authors	Details
	of pentachlorophenol to AML 12 mouse hepatocytes [10].
Xiao and Singh [11]	Xiao and Singh discussed on diallyl trisulfide, a constituent of processed garlic, inactivating Akt to trigger mitochondrial translocation of BAD and caspase-mediated apoptosis in human prostate cancer cells [11].
Fukamachi et al [12]	Fukamachi et al reported on possible enhancing effects of atrazine and nonylphenol on 7,12-dimethylbenz[a]anthracene-induced mammary tumor development in human c-Ha-ras proto-oncogene transgenic rats [12].
Srivastava et al [13]	Srivastava et al reported on toxicological effects of malachite green [13].
Jaga and Duvvi [14]	Jaga and Duvvi reported on risk reduction for DDT toxicity and carcinogenesis through dietary modification [14].
Sundarrajan et al [15]	Sundarrajan et al reported on overexpression of G1/S cyclins and PCNA and their relationship to tyrosine phosphorylation and dephosphorylation during tumor promotion by metanil yellow and malachite green [15].
Alavanja et al [16]	Alavanja et al discussed on the concept "The Agricultural Health Study" [16].
Djordjevic et al [17]	Djordjevic et al reported on an assessment of chlorinated pesticide residues in cigarette tobacco based on supercritical fluid extraction and GC-ECD [17].
Hard et al [18]	Hard et al discussed on identity and pathogenesis of stomach tumors in Sprague-Dawley rats associated with the dietary administration of butachlor [18].
Beier [19]	Beier discussed on natural pesticides and bioactive components in foods [19].
Sparnins et al [20]	Sparnins et al reported on effects of organosulfur compounds from garlic and onions on benzo[a]pyrene-induced neoplasia and glutathione S-transferase activity in the mouse [20].
Gosálvez [21]	Gosálvez discussed on carcinogenesis with the insecticide rotenone [21].

REFERENCES

[1] Cho WS, Han BS, Nam KT, Park K, Choi M, Kim SH, Jeong J, Jang DD. Carcinogenicity study of 3-monochloropropane-1,2-diol in Sprague-Dawley rats.*Food Chem. Toxicol.* 2008 Sep;46(9):3172-7.

[2] Cho WS, Han BS, Lee H, Kim C, Nam KT, Park K, Choi M, Kim SJ, Kim SH, Jeong J, Jang DD. Subchronic toxicity study of 3-monochloropropane-1,2-diol administered by drinking water to B6C3F1 mice. *Food Chem. Toxicol.* 2008 May;46(5):1666-73.

[3] Kuhn K, Nowak B, Klein G, Behnke A, Seidel A, Lampen A. Determination of polycyclic aromatic hydrocarbons in smoked pork by effect-directed bioassay with confirmation by chemical analysis. *J. Food Prot.* 2008 May;71(5):993-9.

[4] Bellocq D, Molina J, Rathahao E, Canlet C, Taché S, Martin PG, Pierre F, Paris A. High potency of bioactivation of 2-amino-1-methyl-6-phenylimidazo[4,5-b]pyridine (PhIP) in mouse colon epithelial cells with Apc(Min) mutation. *Mutat Res.* 2008 May 31;653(1-2):34-43.

[5] Ye L, Leung LK. Effect of dioxin exposure on aromatase expression in ovariectomized rats. *Toxicol. Appl. Pharmacol.* 2008 May 15;229(1):102-8.

[6] Belpomme D, Irigaray P, Hardell L, Clapp R, Montagnier L, Epstein S, Sasco AJ. The multitude and diversity of environmental carcinogens. *Environ. Res.* 2007 Nov;105(3):414-29.

[7] Oesch-Bartlomowicz B, Huelster A, Wiss O, Antoniou-Lipfert P, Dietrich C, Arand M, Weiss C, Bockamp E, Oesch F. Aryl hydrocarbon receptor activation by cAMP vs. dioxin: divergent signaling pathways. *Proc. Natl. Acad Sci. U S A.* 2005 Jun 28;102(26):9218-23.

[8] Hardell L, Lindström G, Liljegren G, Dahl P, Magnuson A. Increased concentrations of octachlorodibenzo-p-dioxin in cases with breast cancer--results from a case-control study. *Eur. J.Cancer Prev.* 1996 Oct;5(5):351-7.

[9] Wong RH, Chang SY, Ho SW, Huang PL, Liu YJ, Chen YC, Yeh YH, Lee HS. Polymorphisms in metabolic GSTP1 and DNA-repair XRCC1 genes with an increased risk of DNA damage in pesticide-exposed fruit growers. *Mutat Res.* 2008 Jul 31;654(2):168-75.

[10] Dorsey WC, Tchounwou PB, Sutton D. Mitogenic and cytotoxic effects of pentachlorophenol to AML 12 mouse hepatocytes. *Int. J. Environ. Res. Public Health.* 2004 Sep;1(2):100-5.

[11] Xiao D, Singh SV. Diallyl trisulfide, a constituent of processed garlic, inactivates Akt to trigger mitochondrial translocation of BAD and caspase-

mediated apoptosis in human prostate cancer cells. *Carcinogenesis.* 2006 Mar;27(3):533-40.

[12] Fukamachi K, Han BS, Kim CK, Takasuka N, Matsuoka Y, Matsuda E, Yamasaki T, Tsuda H. Possible enhancing effects of atrazine and nonylphenol on 7,12-dimethylbenz[a]anthracene-induced mammary tumor development in human c-Ha-ras proto-oncogene transgenic rats. *Cancer Sci.* 2004 May;95(5):404-10.

[13] Srivastava S, Sinha R, Roy D. Toxicological effects of malachite green. Aquat Toxicol. 2004 Feb 25;66(3):319-29.

[14] Jaga K, Duvvi H. Risk reduction for DDT toxicity and carcinogenesis through dietary modification. *J. R. Soc. Health.* 2001 Jun;121(2):107-13.

[15] Sundarrajan M, Fernandis AZ, Subrahmanyam G, Prabhudesai S, Krishnamurthy SC, Rao KV. Overexpression of G1/S cyclins and PCNA and their relationship to tyrosine phosphorylation and dephosphorylation during tumor promotion by metanil yellow and malachite green. *Toxicol. Lett.* 2000 Jul 27;116(1-2):119-30.

[16] Alavanja MC, Sandler DP, McMaster SB, Zahm SH, McDonnell CJ, Lynch CF, Pennybacker M, Rothman N, Dosemeci M, Bond AE, Blair A. The Agricultural Health Study. *Environ. Health Perspect.* 1996 Apr;104(4):362-9.

[17] Djordjevic MV, Fan J, Hoffmann D. Assessment of chlorinated pesticide residues in cigarette tobacco based on supercritical fluid extraction and GC-ECD. *Carcinogenesis.* 1995 Nov;16(11):2627-32.

[18] Hard GC, Iatropoulos MJ, Thake DC, Wheeler D, Tatematsu M, Hagiwara A, Williams GM, Wilson AG. Identity and pathogenesis of stomach tumors in Sprague-Dawley rats associated with the dietary administration of butachlor. *Exp. Toxicol. Pathol.* 1995 May;47(2-3):95-105.

[19] Beier RC. Natural pesticides and bioactive components in foods. *Rev. Environ. Contam. Toxicol.* 1990;113:47-137.

[20] Sparnins VL, Barany G, Wattenberg LW. Effects of organosulfur compounds from garlic and onions on benzo[a]pyrene-induced neoplasia and glutathione S-transferase activity in the mouse. *Carcinogenesis.* 1988 Jan;9(1):131-4.

[21] Gosálvez M. Carcinogenesis with the insecticide rotenone. *Life Sci.* 1983 Feb 21;32(8):809-16.

Chapter 11

ALCOHOLIC BEVERAGE

INTRODUCTION

Alcohol is one of the well-known liquid for human beings. Alcoholic beverage or liquor has been drunk by human beings for a long time. This is usually for social purpose. Alcohol has very high energy but is not useful for health. There are several kinds of alcohol but the only one that can be drunk without danger is ethanol. Around the world, there are several alcoholic beverage manufacturers. The alcoholic drinking can cause several problems. Some countries ban selling of alcoholic beverage due to the religious regulations.

Consumption of large amount of beer, wine and liquor, especially in smokers, increases risks for gastrointestinal tract cancers. The quoted cancers include stomach cancer, pharyngeal cancer, esophageal cancer, liver cancer, stomach cancer, pancrease cancer as well as colorectal cancer. Indeed, the ethanol has no carcinogenic property. But it has several effects on metabolic system. This can lead to cancer. Alcohol promote metabolism of many carcinogens. Alocohol also destroys normal regulative metabolism. Alcohol also leads to malnutrition which results in impairment of immunity (including cancer immunity). These are the facts that alcohol results in cancer. In addition, illegal alcoholic beverage can be seen in many developing countries and these products are harmful to the consumers. Local food producers in those countries produce illegal alcoholic beverages by adding some illegal substance. For sure, this is not a standard protocol. The quoted problematic substances include methanol, which can result in blindness of the consumers, pesticide and washing powder. Methanol is the most widely mentioned contamination. This brings blindness as already mentioned and must be controlled. The antidote for this substance is ethanol, the consumable alcohol. There are many reports on the methanol intoxication due to ingestion of illegal alcoholic beverages contaminated with methanol.

ALCOHOL CONSUMPTION AND CARCINOGENESIS

The topic whether alcohol consumption relates to carcinogenesis is widely studied for a long times. A number of researches on this area are done and published. The important ones will be quoted in Table 1.

Table 1. Important reports on alcohol consumption and carcinogenesis

Authors	Details
Qiu et al [1]	Qiu et al reported on synergistic effect of HBV infection, alcohol and raw fish consumption on oncogenisis of primary hepatic carcinoma [1].
Jelski and Szmitkowski [2]	Jelski and Szmitkowski discussed on alcohol dehydrogenase and aldehyde dehydrogenase in the cancer diseases [2].
Lee et al [3]	Lee et al reported on allelic variants of cytochrome P4501A1 (CYP1A1), glutathione S transferase M1 (GSTM1) polymorphisms and their association with smoking and alcohol consumption as gastric cancer susceptibility biomarkers [3].
Mason and Choi [4]	Mason and Choi discussed on effects of alcohol on folate metabolism and further implied for carcinogenesis [4].
Salaspuro and Salaspuro [5]	Salaspuro and Salaspuro reported on synergistic effect of alcohol drinking and smoking on in vivo acetaldehyde concentration in saliva [5].
Pöschl and Seitz [6]	Pöschl and Seitz discussed on the topic alcohol and cancer [6].
Liao et al [7]	Liao et al reported on lack of correlation of betel nut chewing, tobacco smoking, and alcohol consumption with telomerase activity and the severity of oral cancer [7].
Gallus et al [8]	Gallus et al discussed on laryngeal cancer in women focusing on tobacco, alcohol, nutritional, and hormonal factors [8].

Authors	Details
Cai et al [9]	Cai et al reported on risk factors for the gastric cardia cancer in a case-control study in Fujian Province, China [9]. According to this work, alcoholic beverage consumption is an important risk factor [9].
Muto et al [10]	Muto et al repored on an association between aldehyde dehydrogenase gene polymorphisms and the phenomenon of field cancerization in patients with head and neck cancer [10].
Simanowski [11]	Simanowski reported on increased rectal cell proliferation following alcohol abuse [11].
Löw-Baselli et al [12]	Löw-Baselli et al reported on failure to demonstrate chemoprevention by the monoterpene perillyl alcohol during early rat hepatocarcinogenesis [12].
Seitz et al [13]	Seitz et al reported on cell proliferation and its evaluation in the colorectal mucosa focusing on effect of ethanol [12].
Seitz et al [14] Wada et al [15]	Seitz et al discussed on alcohol and cancer [14]. Wada et al rported on effects of catechol, sodium chloride and ethanol either alone or in combination on gastric carcinogenesis in rats pretreated with N-methyl-N'-nitro-N-nitrosoguanidine [15].
Glynn et al [16]	Glynn et al discussed on alcohol consumption and risk of colorectal cancer in a cohort of Finnish men [16].
Cerar and Pokorn [17]	Cerar and Pokorn reported on inhibition of MNNG-induced gastroduodenal carcinoma in rats by synchronous application of wine or 11% ethanol [17].
Simanowski et al [18]	Simanowski et al reported on effect of alcohol on gastrointestinal cell regeneration as a possible mechanism in alcohol-associated carcinogenesis [18].
Castelletto et al [19]	Castelletto et al discussed on alcohol, tobacco, diet, mate drinking, and esophageal cancer in Argentina [19].
Simanowski et al [20]	Simanowski et al reported on enhancement of ethanol induced rectal mucosal hyper regeneration with age in F344 rats [20].
al-Damouk [21]	al-Damouk reported on oral epithelial response to experimental chronic alcohol ingestion in hamsters [21].

Authors	Details
Yonekura et al [22]	Yonekura et al reported that ethanol ingestion combined with lowered carbohydrate intake could enhance the initiation of diethylnitrosamine liver carcinogenesis in rats [22].
Yamagiwa et al [23]	Yamagiwa et al reported on effect of alcohol ingestion on carcinogenesis by synthetic estrogen and progestin in the rat liver [23].
Seitz et al [24]	Seitz et al reported on possible role of acetaldehyde in ethanol-related rectal cocarcinogenesis in the rat [24].
Woutersen et al [25]	Woutersen et al reported on modulation of dietary fat-promoted pancreatic carcinogenesis in rats and hamsters by chronic ethanol ingestion [25].
Rogers and Conner [26]	Rogers and Conner discussed on alcohol and cancer [26].
Seitz et al [27]	Seitz et al reported on ethanol and intestinal carcinogenesis in the rat [27].
Seitz et al [28]	Seitz et al reported on stimulation of chemically induced rectal carcinogenesis by chronic ethanol ingestion [28].
Seitz et al [29]	Seitz et al reported on enhancement of 1,2-dimethylhydrazine-induced rectal carcinogenesis following chronic ethanol consumption in the rat [29].
Radike et al [30]	Radike et al reported on effect of ethanol on vinyl chloride carcinogenesis [30].

REFERENCES

[1] Qiu XQ, Tan SK, Yu HP, Zeng XY, Li LQ, Hu L. Synergistic effect of HBV infection, alcohol and raw fish consumption on oncogenisis of primary hepatic carcinoma. Zhonghua Zhong Liu Za Zhi. 2008 Feb;30(2):113-5.

[2] Jelski W, Szmitkowski M. Alcohol dehydrogenase (ADH) and aldehyde dehydrogenase (ALDH) in the cancer diseases. Clin. Chim. Acta. 2008 Sep;395(1-2):1-5.

[3] Lee K, Cáceres D, Varela N, Csendes D A, Ríos R H, Quiñones S L. [Allelic variants of cytochrome P4501A1 (CYP1A1), glutathione S transferase M1 (GSTM1) polymorphisms and their association with

smoking and alcohol consumption as gastric cancer susceptibility biomarkers. *Rev. Med. Chil.* 2006 Sep;134(9):1107-15.

[4] Mason JB, Choi SW. Effects of alcohol on folate metabolism: implications for carcinogenesis. *Alcohol.* 2005 Apr;35(3):235-41.

[5] Salaspuro V, Salaspuro M. Synergistic effect of alcohol drinking and smoking on in vivo acetaldehyde concentration in saliva. *Int. J. Cancer.* 2004 Sep 10;111(4):480-3.

[6] Pöschl G, Seitz HK. Alcohol and cancer. *Alcohol. Alcohol.* 2004 May-Jun;39(3):155-65.

[7] Liao CT, Chen IH, Chang JT, Wang HM, Hsieh LL, Cheng AJ. Lack of correlation of betel nut chewing, tobacco smoking, and alcohol consumption with telomerase activity and the severity of oral cancer. *Chang. Gung. Med. J.* 2003 Sep;26(9):637-45.

[8] Gallus S, Bosetti C, Franceschi S, Levi F, Negri E, La Vecchia C. Laryngeal cancer in women: tobacco, alcohol, nutritional, and hormonal factors. *Cancer Epidemiol. Biomarkers Prev.* 2003 Jun;12(6):514-7.

[9] Cai L, Zheng ZL, Zhang ZF. Risk factors for the gastric cardia cancer: a case-control study in Fujian Province. *World J. Gastroenterol.* 2003 Feb;9(2):214-8.

[10] Muto M, Nakane M, Hitomi Y, Yoshida S, Sasaki S, Ohtsu A, Yoshida S, Ebihara S, Esumi H. Association between aldehyde dehydrogenase gene polymorphisms and the phenomenon of field cancerization in patients with head and neck cancer. *Carcinogenesis.* 2002 Oct;23(10):1759-65.

[11] Simanowski UA, Homann N, Knühl M, Arce L, Waldherr R, Conradt C, Bosch FX, Seitz HK. Increased rectal cell proliferation following alcohol abuse. *Gut.* 2001 Sep;49(3):418-22.

[12] Löw-Baselli A, Huber WW, Käfer M, Bukowska K, Schulte-Hermann R, Grasl-Kraupp B. Failure to demonstrate chemoprevention by the monoterpene perillyl alcohol during early rat hepatocarcinogenesis: a cautionary note. *Carcinogenesis.* 2000 Oct;21(10):1869-77.

[13] Seitz HK, Simanowski UA, Homann N, Waldherr R. Cell proliferation and its evaluation in the colorectal mucosa: effect of ethanol. *Z Gastroenterol.* 1998 Aug;36(8):645-55.

[14] Seitz HK, Pöschl G, Simanowski UA. Alcohol and cancer. *Recent Dev. Alcohol.* 1998;14:67-95.

[15] Wada S, Hirose M, Shichino Y, Ozaki K, Hoshiya T, Kato K, Shirai T. Effects of catechol, sodium chloride and ethanol either alone or in combination on gastric carcinogenesis in rats pretreated with N-methyl-N'-nitro-N-nitrosoguanidine. *Cancer Lett.* 1998 Jan 30;123(2):127-34.

[16] Glynn SA, Albanes D, Pietinen P, Brown CC, Rautalahti M, Tangrea JA, Taylor PR, Virtamo J. Alcohol consumption and risk of colorectal cancer in a cohort of Finnish men. *Cancer Causes Control.* 1996 Mar;7(2):214-23.
[17] Cerar A, Pokorn D. Inhibition of MNNG-induced gastroduodenal carcinoma in rats by synchronous application of wine or 11% ethanol. *Nutr. Cancer.* 1996;26(3):347-52.
[18] Simanowski UA, Stickel F, Maier H, Gärtner U, Seitz HK. Effect of alcohol on gastrointestinal cell regeneration as a possible mechanism in alcohol-associated carcinogenesis. *Alcohol.* 1995 Mar-Apr;12(2):111-5.
[19] Castelletto R, Castellsague X, Muñoz N, Iscovich J, Chopita N, Jmelnitsky A. Alcohol, tobacco, diet, mate drinking, and esophageal cancer in Argentina. *Cancer Epidemiol. Biomarkers Prev.* 1994 Oct-Nov;3(7):557-64.
[20] Simanowski UA, Suter P, Russell RM, Heller M, Waldherr R, Ward R, Peters TJ, Smith D, Seitz HK. Enhancement of ethanol induced rectal mucosal hyper regeneration with age in F344 rats. *Gut.* 1994 Aug;35(8):1102-6.
[21] al-Damouk JD. Oral epithelial response to experimental chronic alcohol ingestion in hamsters. *Oral Surg. Oral. Med. Oral. Pathol.* 1993 Dec;76(6):736-41.
[22] Yonekura I, Matsumoto Y, Miura K, Sato A. Ethanol ingestion combined with lowered carbohydrate intake enhances the initiation of diethylnitrosamine liver carcinogenesis in rats. *Nutr. Cancer.* 1992;17(2):171-8.
[23] Yamagiwa K, Higashi S, Mizumoto R. Effect of alcohol ingestion on carcinogenesis by synthetic estrogen and progestin in the rat liver. *Jpn. J. Cancer Res.* 1991 Jul;82(7):771-8.
[24] Seitz HK, Simanowski UA, Garzon FT, Rideout JM, Peters TJ, Koch A, Berger MR, Einecke H, Maiwald M. Possible role of acetaldehyde in ethanol-related rectal cocarcinogenesis in the rat. *Gastroenterology.* 1990 Feb;98(2):406-13.
[25] Woutersen RA, van Garderen-Hoetmer A, Bax J, Scherer E. Modulation of dietary fat-promoted pancreatic carcinogenesis in rats and hamsters by chronic ethanol ingestion. *Carcinogenesis.* 1989 Mar;10(3):453-9.
[26] Rogers AE, Conner MW. Alcohol and cancer. Adv Exp Med Biol. 1986;206:473-95.
[27] Seitz HK, Czygan P, Waldherr R, Veith S, Kommerell B. Ethanol and intestinal carcinogenesis in the rat. *Alcohol.* 1985 May-Jun;2(3):491-4.

[28] Seitz HK, Czygan P, Simanowski U, Waldherr R, Veith S, Raedsch R, Kommerell B. Stimulation of chemically induced rectal carcinogenesis by chronic ethanol ingestion. *Alcohol Alcohol.* 1985;20(4):427-33.

[29] Seitz HK, Czygan P, Waldherr R, Veith S, Raedsch R, Kässmodel H, Kommerell B. Enhancement of 1,2-dimethylhydrazine-induced rectal carcinogenesis following chronic ethanol consumption in the rat. Gastroenterology. 1984 May;86(5 Pt 1):886-91.

[30] Radike MJ, Stemmer KL, Bingham E. Effect of ethanol on vinyl chloride carcinogenesis. *Environ. Health Perspect.* 1981 Oct;41:59-62.

INDEX

A

acetaldehyde, 100, 103, 104
acetaminophen, 40, 45
acetic acid, 12, 13
acid, 12, 13, 14, 15, 18, 19, 22, 25, 26, 28, 31, 35, 39, 40, 45, 46, 53, 56, 58, 59, 60, 76, 79, 80, 82, 84
aCL, 81
acquired immunodeficiency syndrome, 2
activation, 40, 45, 71, 72, 95, 97
acute, 14, 15, 21, 26, 31
acute renal failure, 14
Adams, 6
additives, 76, 78, 80, 87, 88
adducts, 27, 33, 40, 46, 64, 70, 90
adenoma, 50, 51, 52, 56, 76, 79
adenomas, 56
ADH, 102
adiposity, 6
administration, 95, 98
adolescents, 6, 7
adsorption, 22, 29
adult, 12, 63, 79, 88
adult population, 88
adulteration, 19
adults, 6, 88
aerobic, 22, 29
Afghanistan, 6
aflatoxins, 22, 25, 26, 27, 28, 30, 31, 33

Africa, 7, 21, 27, 33, 57
age, 6, 27, 33, 57, 100, 104
agent, 11, 31
agents, 8, 9, 15, 19, 27, 32, 36, 43, 50
aging, 22, 29
agricultural, 36, 43, 93
agricultural crop, 36, 43
agriculture, 5
AIDS, 2
air, 88, 92, 93
albumin, 27, 33
alcohol, 53, 59, 99, 100, 102, 103, 104
alcohol abuse, 100, 103
alcohol consumption, 100, 103
alleles, 56
alpha, 49, 53, 58, 65, 66, 71, 82, 84
alternative, 89
American Cancer Society, 84
amine, 40, 42, 45, 76, 79, 80, 82
amines, 40, 42, 45, 90
amino, 35, 39, 42, 43, 95, 97
amino acid, 35, 39
amino acids, 39
AML, 95, 97
anaemia, 21
anaerobic, 78
anaerobic bacteria, 78
animal models, 14, 35, 39, 42
animals, 21, 32, 64, 70, 90, 92, 93, 94
anthracene, 95, 98
antibacterial, 53, 58

antibiotic, 87
antibody, 27, 33
anticancer, 53, 58
anti-HER2, 53
antioxidant, 57, 76, 79
antitumor, 50
anti-tumor, 49
Anti-tumor, 55
anus, 61
apoptosis, 50, 52, 53, 58, 71, 95, 98
appetite, 3, 82
apples, 22, 29
application, 53, 59, 95, 100, 104
Arabia, 22, 29
Argentina, 100, 104
aromatic hydrocarbons, 36, 43, 44, 45
arrest, 50
arsenic, 9
arsenic poisoning, 9
aryl hydrocarbon receptor, 95
ascorbic, 76, 79, 80, 82, 84
ascorbic acid, 76, 79, 80, 82, 84
Asia, 5, 11, 17, 56, 61, 63, 65, 75
Asian, 3, 5, 31, 68, 69, 70, 71, 75
Asian countries, 5, 75
asparagines, 39
aspirate, 53, 59
assessment, 47, 95
atherosclerosis, 57
atmosphere, 24
ATP, 71
attitudes, 8
avoidance, 76

B

babies, 14, 18
bacteria, 22, 28, 78
bacterial, 2, 3, 4, 22, 29
bacterial strains, 22, 29
barley, 21, 22, 29
Bax, 104
beef, 8, 42
beer, 99
behavior, 3, 52, 67

Belgium, 7
benzo(a)pyrene, 36, 40, 44, 45
beverage manufacturers, 99
beverages, 13, 19, 99
bicarbonate, 82, 84
bile, 53, 82, 84
biliary tract, 61, 63
binding, 36, 44
bioassay, 15, 36, 43, 95, 97
bioavailability, 36
biological processes, 51
biomarker, 90, 92
biomarkers, 6, 40, 46, 100, 103
biosecurity, 8
biota, 36, 45
bird flu, 17
birth, 53, 59
birth weight, 53, 59
bladder, 15, 19, 73, 76, 80, 82, 84, 88, 89, 92
bladder cancer, 15, 76, 80, 89
bladder carcinogenesis, 19, 88, 92
bladder stones, 15, 19
bleeding, 51
blindness, 99
blood, 6, 22, 28, 50
blood glucose, 6
blood vessels, 50
BMI, 6
body mass index, 7, 53, 59
body size, 7
body weight, 12, 53, 58
boiling, 35, 39
bone density, 56
boric acid, 26, 31
botulinum, 4
botulism, 3
Brazil, 5
breast cancer, 39, 42, 52, 53, 58, 95, 97
breastfeeding, 53, 59
broilers, 22, 28

C

Ca^{2+}, 71, 72
caffeine, 82, 84, 88, 92

Index

calcium, 19, 50, 51, 56, 72
calcium oxalate, 19
calculus, 15, 19
caloric intake, 7
calorie, 6
cAMP, 95, 97
Canada, 79, 92
cancer, 4, 6, 9, 12, 27, 32, 36, 39, 40, 42, 43, 45, 46, 52, 53, 58, 59, 60, 61, 63, 65, 76, 78, 79, 80, 81, 82, 83, 84, 89, 93, 94, 95, 99, 100, 102, 103, 104
cancer cells, 53, 58, 95, 98
candida, 24, 30
capillary, 20, 78
carbohydrate, 1, 52, 57, 89, 100, 104
carcinogen, 15, 20, 40, 42, 43, 46, 65, 76
carcinogenesis, ix, 2, 4, 9, 15, 19, 20, 26, 27, 32, 33, 36, 39, 40, 42, 53, 56, 63, 64, 66, 67, 70, 71, 75, 76, 78, 80, 81, 82, 84, 85, 88, 89, 90, 92, 94, 95, 98, 100, 103, 104, 105
carcinogenic, 25, 26, 30, 35, 36, 38, 39, 42, 45, 71, 90, 92, 99
carcinogenicity, 19, 40, 42, 47, 89, 90, 94
carcinogens, 9, 35, 36, 42, 43, 78, 90, 92, 95, 97, 99
carcinoma, 25, 26, 27, 30, 32, 33, 49, 50, 51, 52, 64, 70, 73, 79, 100, 102, 104
carcinomas, 55, 73
cardiovascular disease, 51, 56, 57
cardiovascular system, 52
carrier, 3
CAS, 20, 90, 91
caspase, 95, 97
catabolism, 49
catalyst, 76
catechol, 82, 84, 100, 103
cats, 14, 18, 19
causal relationship, 26
cell, 22, 29, 51, 52, 53, 58, 71, 72, 88, 90, 91, 92, 100, 103, 104
cell cycle, 52, 53, 58, 88, 91
cell death, 53, 58
cell line, 71
certification, 16

cheese, 36, 45
chemicals, 1
chemometrics, 20
chemoprevention, 100, 103
chewing, 100, 103
Chi square, 50
chicken, 17
childhood, 6
children, 6, 7, 8, 36, 43, 79, 88
Chile, 5
China, ix, 4, 14, 17, 18, 25, 27, 30, 32, 64, 70, 76, 80, 100
chloride, 81, 82, 84, 100, 103, 105
cholangiocarcinoma, 4, 63, 64, 65, 67, 69, 70, 71, 76
cholecalciferol, 49
cholera, 2, 3
cholesterol, 57
chronic diseases, 56, 57
chronic irritation, 63
cigarette smoking, 36, 44
cirrhosis, 27
citizens, 18
classical, 81, 90
classical conditioning, 90
cleaning, 13, 22, 29
coagulation, 32
coffee, 22, 28
cohort, 40, 46, 100, 104
collaboration, 17
collagen, 36, 44
colon, 49, 51, 52, 56, 76, 80, 82, 84, 95, 97
colon cancer, 52, 56, 76, 80, 82, 84
colon carcinogenesis, 56
colon polyps, 51
colorectal cancer, 36, 43, 49, 56, 99, 100, 104
colors, 87
combined effect, 90
components, 95, 98
composition, 11, 12
compounds, 36, 40, 43, 45, 57, 76, 79, 90, 92, 95, 98
concentration, 12, 76, 79, 80, 100, 103
conditioning, 92
consolidation, 8

consumers, 11, 21, 87, 99
consumption, 1, 36, 42, 44, 45, 53, 100, 102, 103, 104, 105
contaminant, ix, 2, 3, 4, 11, 23, 24, 35, 36, 38, 39
contaminants, 3, 4, 21, 35, 36, 43
contaminated food, 16
contamination, ix, 1, 2, 3, 4, 8, 11, 12, 13, 15, 16, 17, 18, 21, 22, 24, 25, 26, 30, 31, 35, 36, 39, 40, 42, 45, 62, 88, 93, 94, 95, 99
control, 16, 17, 26, 31, 49, 51, 64, 70, 71, 82, 84, 97, 100, 103
cooking, 24, 39, 61
corn, 22, 25, 29, 30, 31, 53, 57
coronary heart disease, 57
correlation, 12, 19, 51, 65, 67, 76, 94, 100, 103
correlation coefficient, 12
correlations, 53
counseling, 53, 59
couples, 71
covering, 27
cows, 22, 28
crops, 36, 43
cross-sectional, 6
cross-sectional study, 6
cryptosporidium, 24
crystals, 14, 19
cultivation, 20
culture, 2, 40, 46, 67
CVD, 57
cyclins, 88, 91, 95, 98
cytochrome, 36, 44, 100, 102
cytokine, 52
cytotoxic, 50, 95, 97
cytotoxic agents, 50

D

dairy, 12, 16, 17, 21, 22, 28, 30
dairy products, 12, 16, 17
danger, 38, 39, 99
death, 9, 14, 25, 30, 52, 63
death rate, 52, 63
deaths, 36

defense, 8
deficiency, 1, 81
deficits, 49, 56
degradation, 52
dehydrogenase, 100, 102, 103
density, 52, 56
deoxynivalenol, 22, 28
dephosphorylation, 88, 91, 95, 98
deposition, 14
derivatives, 36, 44
detection, 36, 62, 78
detoxification, 42
developed countries, 1, 7, 8, 81, 93
developing countries, 1, 7, 32, 81, 87, 88, 99
deviation, 2
diabetes, 51, 56
diabetic patients, 89
diarrhea, 2, 21, 24, 30, 35
diarrhoea, 7, 8
diet, 26, 36, 40, 45, 46, 53, 57, 58, 59, 60, 78, 79, 82, 85, 100, 104
dietary, 6, 36, 40, 45, 46, 51, 53, 56, 57, 58, 59, 60, 76, 78, 79, 80, 82, 95, 98, 100, 104
dietary fat, 53, 58, 100, 104
dietary fiber, 57
dietary intake, 36, 40, 46, 76, 79
diets, 53, 57
differentiation, 50, 51, 52, 53, 58
dioxin, 93, 95, 97
direct action, 52
diseases, 64, 70, 100
disinfection, 22, 29
disorder, 1, 52, 61
disseminated intravascular coagulation, 26
distribution, 25, 31, 62, 68, 76, 79, 90
diversity, 9, 36, 43, 62, 95, 97
DNA, 27, 33, 36, 44, 64, 70, 90, 92, 95, 97
DNA damage, 27, 33, 64, 70, 90, 92, 95, 97
dogs, 14, 19, 26, 32
dosage, 42
dosimetry, 27, 32, 40, 46
drinking, 76, 80, 82, 84, 94, 97, 99, 100, 103, 104
drinking water, 76, 80, 82, 84, 94, 97
drug use, 56

dyes, 88, 92
dysregulated, 88, 91

E

E.coli, 3
East Asia, 11, 17
eating, 3, 35, 65, 67, 76
eating behavior, 67
egg, 17
electrochemical detection, 78
electrolytes, 81
electrophoresis, 20, 78
elephant, 57
Endocannabinoid system, 56
endocrine, 50, 51
endocrine system, 51
endometrial carcinoma, 52
endothelium, 50
energy, 1, 6, 7, 52, 89, 99
enterovirus, 3
environment, 7, 9, 21, 93
environmental contaminants, 4, 36, 43
enzyme induction, 27, 32
enzymes, 27, 32
EPA, 20
epidemiology, 40, 46
epidermal growth factor, 65
epigenetic, 64, 70
epigenetic alterations, 64, 70
epithelial cell, 95, 97
epithelial cells, 95, 97
epithelium, 52
Escherichia coli, 22, 29
esophageal cancer, 76, 82, 99, 100, 104
ester, 56
estimating, 76, 80
Estonia, 36, 43
estrogen, 49, 53, 59, 100, 104
ethanol, 82, 84, 99, 100, 103, 104, 105
etiology, 9, 27, 32, 76, 80
eukaryotes, 56
Euro, 18
European Union, 18
excretion, 40, 46, 64, 70, 89

exercise, 6, 56
exocytosis, 71
exposure, 15, 27, 32, 33, 35, 36, 40, 42, 44, 45, 46, 65, 66, 68, 71, 76, 80, 90, 95, 97
extra virgin oliv, 53, 57
extraction, 22

F

failure, 14, 100
false positive, 11
FAO, 43
fasting, 40, 46
fat, 1, 52, 53, 56, 58, 59, 76, 78, 80
fats, 57
fatty acids, 53, 57, 59, 60
FDA, 17, 18
fear, 3, 12
feeding, 12, 19, 53, 60
fermentation, 22, 28
fertilizer, 11
fiber, 76, 78, 80
fibers, 57
fibrosis, 72
fire, 11
fish, 19, 36, 44, 62, 65, 67, 68, 75, 82, 100, 102
flame, 39
fluid, 53, 59, 95, 98
fluid extract, 95, 98
focusing, 25, 27, 40, 63, 64, 66, 67, 75, 76, 100
folate, 100, 103
food, ix, 1, 2, 3, 4, 5, 6, 8, 11, 12, 13, 16, 17, 18, 19, 21, 22, 23, 24, 25, 30, 31, 35, 36, 38, 39, 40, 42, 43, 44, 45, 46, 47, 52, 53, 56, 65, 71, 75, 76, 78, 79, 80, 81, 82, 87, 88, 89, 90, 93, 95, 99
food additives, 76, 78, 80, 88
food intake, 56
food production, 4
food products, 2, 31, 42, 76, 79
food safety, ix, 2, 4, 8, 16, 87
foodstuffs, 78
formaldehyde, 12, 13, 18, 19

Fox, 79, 80, 83, 92
France, 9, 72
free radical, 53, 76
free radicals, 53
fresh water, 61, 65
freshwater, 61, 62, 68
fuel, 57
fungal, 4, 21, 22, 24
fungi, 22, 28, 29
fungus, 21, 24, 30
Fusarium, 22, 28, 29

G

gas, 53, 59, 60
gastric, 39, 76, 78, 79, 80, 81, 82, 83, 84, 85, 100, 103
gastric mucosa, 76, 80
gastrointestinal, 14, 76, 79, 99, 100, 104
gastrointestinal tract, 14, 76, 99
gender, 40, 46
gene, 27, 32, 33, 50, 52, 53, 58, 78, 100, 103
gene expression, 78
generation, 13, 39, 53, 76
genes, 53, 58, 60, 65, 66, 95, 97
genetic alteration, 73
genetics, 7
Geneva, 43
genistein, 49
genotoxic, 15, 19, 40, 45
glioma, 79
globulin, 82, 84
glucose, 6
glutamate, 3
glutathione, 95, 98, 100, 102
glycol, 18
goiter, 81
grain, 36, 43
groups, 66
growth, 49, 50, 52, 53, 59, 60, 65, 71
growth factor, 50, 52, 65
guanine, 27, 32
Guinea, 22, 28

H

habitat, 61
HBV, 100, 102
HBV infection, 100, 102
head and neck cancer, 100, 103
headache, 75
health, 7, 8, 12, 16, 21, 23, 26, 42, 47, 53, 56, 60, 75, 81, 87, 88, 99
health care, 8
health education, 12
health effects, 21, 75, 87
health problems, 88
heart disease, 56
heart failure, 57
heat, 12, 22, 24, 29, 35, 40, 46
heating, 35, 38, 39, 42
Helicobacter pylori, 82, 83
hemoglobin, 40, 46
hemostasis, 57
hepatic encephalopathy, 26, 31
hepatitis, 26
hepatitis B, 26
hepatocarcinogenesis, 72, 88, 91, 100, 103
hepatocellular, 25, 26, 27, 30, 32, 33, 63, 64, 69
hepatocellular carcinoma, 25, 26, 27, 30, 32, 33, 64
hepatocytes, 95, 97
hepatoma, 71
HER2, 58
herbs, 36, 44
high fat, 53, 60
high temperature, 35, 39
high-fat, 53, 58
high-performance liquid chromatography (HPLC), 36, 45, 78
high-risk, 7, 76, 80
histochemical, 14
Holland, 6, 31
hormone, 51, 53, 59
hospitalizations, 14
host, 61, 62, 65, 68, 69
households, 5
H-ras, 73

HSC, 72
human, 1, 6, 7, 9, 12, 14, 17, 23, 25, 26, 27, 30, 32, 36, 40, 41, 42, 43, 44, 45, 46, 49, 52, 53, 56, 58, 60, 61, 62, 64, 65, 69, 71, 76, 78, 79, 80, 81, 82, 89, 90, 94, 95, 98, 99
human development, 6
human exposure, 36, 43, 45
human milk, 36, 44
humans, 14, 26, 27, 32, 40, 46, 47, 57, 76, 80, 90, 92
hybrids, 25, 31
hydro, 36, 43, 44, 45, 95, 97
hydrocarbon, 35, 36, 38, 39, 40, 42, 44, 45, 95, 97
hydrocarbons, 36
hygiene, 3, 7, 8, 21
hyperinsulinemia, 53, 60
hypermethylation, 64, 70
hypernatremia, 81
hypertension, 81
hypertriglyceridemia, 56
hypothesis, 6
hypovitaminosis D, 50

I

IARC, 79
ice, 40, 47
identification, 14, 22, 28, 31, 36, 44, 64, 70
identity, 95
immigrants, 64, 70
immune response, 51
immune system, 50
immunity, 99
in vitro, 42, 49, 72
in vivo, 40, 46, 49, 72, 76, 88, 91, 100, 103
incidence, 25, 30, 51, 52, 67, 82, 90
India, 8
Indian, 62, 68, 91
Indian Ocean, 62, 68
indicators, 6
indirect effect, 3
inducer, 65
induction, 15, 19, 26, 27, 32, 36, 50, 64, 70, 76, 90, 91, 92

industrial, 93
infants, 11, 14
infection, 35, 62, 63, 65, 68, 82, 100, 102
infectious disease, 3, 24, 30, 61
infestations, 61
inflammation, 63, 64, 70
inflammatory, 56, 64, 70
inflammatory disease, 64, 70
ingest, 2, 12, 61, 88
ingestion, 1, 30, 36, 52, 76, 78, 81, 82, 84, 100, 104, 105
inhibition, 40, 49, 72, 100
inhibitory, 22, 28
initiation, 82, 84, 100, 104
injury, v
insecticide, 17, 93, 95, 98
insecurity, 5
instability, 73
instruments, 53, 59
insults, 26
interaction, 56
interactions, 78
international standards, 16
intervention, 53, 59
intestine, 50
intoxication, 11, 14, 15, 21, 23, 24, 100
intravascular, 26, 32
intron, 52
invasive, 71
iodine, 81
Iran, 25, 31
irritation, 63
isoforms, 72
isolation, 22
isozymes, 36, 44

J

Japan, 82, 84
jaundice, 63, 67
Jun, 5, 6, 7, 8, 18, 19, 29, 30, 32, 43, 44, 45, 46, 47, 56, 58, 59, 68, 69, 70, 71, 72, 79, 80, 83, 84, 91, 97, 98, 103, 104

K

K^+, 71
Kenya, 26, 31
kidney, 14, 50, 89, 94
killing, 4
kinase, 71, 72
Korea, 76, 79

L

laboratory studies, 62, 79
lactic acid, 22, 28
lactic acid bacteria, 22, 28
Lactobacillus, 22, 29
land, 81
Laos, 62, 69
larvae, 61, 65
larval, 62
Laryngeal, 103
laryngeal cancer, 100
Latin America, 6
leptin, 53, 57, 58
lesions, 72, 88, 91
leukemia, 49, 93, 95
life cycle, 62, 65, 68
life style, 6
lifestyle, 36, 56, 65
ligand, 50
lipid, vii, 1, 4, 49, 52, 53, 59, 79
lipid peroxidation, 79
lipid profile, 53, 59
lipids, 53, 59, 60
liquid chromatography, 53, 59, 60, 78
liquor, 99
Listeria monocytogenes, 22, 29
liver, 4, 22, 26, 27, 28, 30, 32, 39, 57, 61, 62, 63, 64, 65, 69, 70, 71, 72, 76, 79, 91, 99, 100, 104
liver cancer, 26, 27, 28, 30, 39, 71, 76, 99
liver disease, 57
London, 32
love, 82, 88
Low cost, 11
lung, 39, 42, 49
lung cancer, 39, 42
lymphoma, 49, 93, 95

M

maize, 22, 26, 28, 31
Malaysia, 26, 31
malignancy, 65
malnutrition, 5, 99
management, 8, 73
manufacturer, 16
manufacturing, 3, 16
MAPK, 72
market, 8, 12, 13, 16, 18
marketing, 11, 16, 88
markets, 22, 29
Markov, 58
mass spectrometry, 20
maternal, 5, 53, 59
measurement, 53, 59, 60
meat, 36, 42, 43, 75, 76, 79, 80
medications, 53, 59
medicine, 1, 2, 3, 21, 23, 26, 76, 87, 94
melamine, ix, 4, 11, 12, 13, 14, 15, 16, 17, 18, 19, 20
men, 53, 57, 59, 100, 104
meta-analysis, 8
metabolic, 56, 57, 95, 97, 99
metabolism, 22, 28, 36, 40, 42, 44, 49, 51, 55, 71, 82, 84, 99, 100, 103
metabolite, 40, 46, 49
methanol, 99
Mexico, 25, 30
mice, 15, 27, 32, 53, 58, 59, 73, 82, 83, 94, 97
microarray, 53, 60
microbes, 21
microbial, 22, 29
micronucleus, 88, 91
micronutrients, 1
microorganisms, 22, 28, 29
microwave, 13
migration, 12, 13, 18
milk, 11, 12, 16, 17, 18, 19, 25, 36, 44, 45, 56
Mississippi, 25, 31

mitochondrial, 95, 97
mitogenic, 95
MMTV, 53, 58
modalities, 6
model system, 22, 29, 49
modeling, 40, 45
models, 14, 35, 39, 42
modulation, 50, 51, 52, 100
MOE, 46
moisture, 24
molecular mass, 36, 45
molecular mechanisms, 53
mollusks, 62, 68, 69
monoclonal, 27, 33
monoclonal antibody, 27, 33
monomers, 13, 18
monounsaturated fat, 53, 58
monounsaturated fatty acids, 53, 58
morphology, 62, 68
mortality, 7, 26, 31, 78, 82, 84
motivation, 8
mouse, 15, 20, 27, 33, 95, 97, 98
mouth, 61
mRNA, 52, 56, 90, 92
mucosa, 100, 103
murder, 9
mushrooms, 22, 29
mutagen, 40, 45
mutagenic, 26, 90, 92
mutation, 95, 97
mutations, 27, 33, 73
myeloma, 49

N

Na^+, 72
N-acety, 64, 70
nationality, 50
natural, 2, 27, 89, 95
nausea, 75
Nebraska, 79
neck, 49, 79, 100, 103
neoplasia, 51, 63, 82, 83, 95, 98
neoplastic, 27, 33, 88, 91
nephropathy, 14

nephrotoxic, 14
nephrotoxicity, 14, 15
network, 66, 67
neuroendocrine, 53, 60
neurotoxic effect, 36
nicotinamide, 72
Nielsen, 28
nitrate, 75, 76, 78, 79, 80
nitrates, 78
nitric oxide, 79
Nitrite, 75, 76, 78
nitrogen, 11
nitrosamines, 78
nitroso compounds, 71, 76, 79, 80
normal, 42, 99
nuclear, 53, 58
nutraceuticals, 53, 58
nutrient, 1, 52
nutrients, 1, 94
nutrition, 1, 5, 6, 7, 56, 78

O

obese, 6, 7
obesity, 6, 7, 53, 57, 59
observations, 91
ODS, 82, 84
oil, 36, 44, 53, 57, 59, 60
oils, 53, 58
old age, 6
olive oil, 53, 57, 59, 60
omega-3, 56
oncogene, 53, 58, 95, 98
oncogenes, 72
oncology, 76
on-line, 20
opioids, 72
optimal health, 56
optimization, 36
oral, 3, 42, 53, 57, 100, 103
oral transmission, 3
organic, 22, 29
organophosphorous, 8
osteoarthritis, 51
ovarian cancer, 52

ovariectomized, 53, 59, 95, 97
ovariectomized rat, 95, 97
overnutrition, 1
over-the-counter, 53, 59
overweight, 5, 6, 7
oxalate, 19
oxidation, 35, 38
oxidative, 22, 28, 64, 90, 92
oxidative stress, 22, 28
oxide, 79
oxygen, 78

P

p53, 27, 32, 33, 72
PAHs, 43, 44
Pakistan, 5
Panama, 25, 31
pancreas, 50, 52
pancreatic, 100, 104
Pap, 43
Papua New Guinea, 22, 28
parasite, 61, 62, 65, 69, 76
parasites, 61
parasitic infestation, 1, 2, 61
parents, 7
Paris, 97
particles, 1
pathogenesis, 64, 70, 76, 95, 98
pathology, 27, 67
pathways, 66, 72
patients, 52, 53, 58, 64, 70, 79, 89, 100, 103
peanuts, 25
pediatric, 1, 24, 30, 88
pelvic, 73
perceptions, 8
perinatal, 27, 33
permeability, 71
personal hygiene, 8
perturbation, 50
pesticide, 93, 95, 97, 98, 99
pesticides, 8, 20, 95, 98
phagocytic, 22, 28
pharmacokinetic, 40, 45
phenolic, 57

phenolic compounds, 57
Philippines, 5
phosphorylation, 88, 91, 95, 98
physical activity, 6
phytoestrogens, 49
pigs, 19
pilot study, 36, 43
PKC, 72
plants, 89, 93
plasma, 22, 29, 53, 59, 60
plastic, 13, 18
poison, 4, 9
poisoning, 9, 24, 26, 29, 31
Poland, 40, 46
poliovirus, 3
pollutant, 93
pollutants, 9, 93
pollution, 39, 93
polycyclic aromatic hydrocarbon, 35, 36, 38, 42, 43, 44, 45, 95, 97
polymorphism, 50, 51, 52, 56
polymorphisms, 42, 50, 52, 56, 64, 70, 95, 100, 102, 103
poor, 3, 4, 5, 11, 63, 64, 70
population, 2, 3, 5, 7, 12, 25, 31, 36, 40, 43, 44, 46, 52, 63, 65, 69, 71, 76, 80, 82, 88
pork, 3, 36, 43, 75, 95, 97
potassium, 22, 29, 82, 84
potato, 40, 46
poultry, 17, 22, 25, 29
poverty, 5
powder, 99
precipitation, 14
pre-clinical, 55
preneoplastic lesions, 72, 88, 91
preservatives, 78
prevention, 7, 9, 55, 56, 79
primary data, 50
primates, 90, 91
producers, 8, 16, 89, 99
production, 11, 16, 17, 22, 25, 27, 28, 31, 33, 50, 52, 81, 82, 87
prognosis, 64, 70
proliferation, 15, 19, 51, 72, 100, 103
promoter, 52, 88, 91

property, v, 15, 35, 38, 39, 99
prostate, 49, 52, 53, 55, 58, 60, 93, 95, 98
prostate cancer, 49, 52, 53, 58, 60, 93, 95, 98
prostate gland, 49
protection, 64, 70
protective factors, 7
protein, 1, 11, 50, 57, 71, 72
protein function, 50
protein kinase C, 71, 72
proteins, 88, 91
protocol, 16, 99
proto-oncogene, 95, 98
public health, 3, 12, 16, 18, 27, 32, 88, 93, 94
public interest, ix, 4
pyrene, 36, 40, 44, 45, 95, 98

Q

query, 2
questionnaire, 46, 76, 79

R

ras, 65, 66, 68, 72, 95, 98
rat, 9, 15, 19, 72, 76, 80, 82, 85, 88, 90, 91, 92, 100, 103, 104, 105
rats, 15, 19, 27, 33, 36, 40, 44, 45, 53, 58, 72, 82, 84, 88, 90, 91, 92, 94, 95, 97, 98, 100, 103, 104
receptors, 50
recovery, 6, 76, 80
recurrence, 51
red meat, 76, 80
redness, 75
redox, 75
regeneration, 100, 104
regression, 53, 59
regular, 16
regulation, 51, 52, 56, 81
regulations, 99
regulators, 20
relationship, 12, 13, 15, 26, 40, 46, 52, 57, 62, 69, 76, 80, 81, 82, 88, 90, 91, 94, 95, 98
relevance, 15, 20, 42, 76, 80, 90

renal, 14, 19, 49, 73
renal failure, 14, 19
repair, 27, 33, 95, 97
reservoir, 62, 69
residential, 36, 44
residues, 36, 44, 95, 98
resin, 12, 13, 18, 19
resistance, 24
respiratory, 39, 93
restriction fragment length polymorphis, 50
risk, 6, 7, 8, 15, 20, 26, 36, 39, 40, 44, 45, 46, 49, 50, 51, 52, 53, 56, 57, 58, 64, 65, 73, 76, 79, 80, 82, 83, 84, 95, 97, 100, 104
risk assessment, 8, 15, 20, 36, 40, 44, 46
risk factors, 8, 57, 64, 73, 82, 83, 100
risks, 8, 42, 99
rodent, 88, 91

S

saccharin, 87, 88, 89, 90, 92
Saccharomyces cerevisiae, 24, 30
safety, 12, 17, 36, 87, 89
saliva, 100, 103
salmon, 36, 43
Salmonella, 3
salt, 75, 76, 81, 82, 83, 84, 85
sample, 40, 46
sanitation, 7, 8
saturated fat, 57
Saudi Arabia, 22, 29
scandal, 17, 18
school, 8
search, 6, 50, 66
search engine, 50
searching, 50
secondary data, 7
security, 5, 8
self-assembly, 18
self-report, 40, 46
SEM, 14
sensing, 50
sensitivity, 53, 59
serum, 27, 32, 33, 52, 53, 57, 58, 90, 92
serum albumin, 27, 33

services, v
severity, 100, 103
SGOT, 26
SGPT, 26
shape, 61
sheep, 21, 30
Shigella, 3
sign, 67
signal transduction, 50, 65, 66
signaling, 50, 52, 53, 58, 97
signaling pathways, 53, 58, 97
signalling, 50
Singapore, 6
skin, 15, 20, 39, 42, 93, 95
skin cancer, 39, 42, 93, 95
small intestine, 78
smoke, 35, 36, 44, 82, 84
smokers, 99
smoking, 35, 36, 40, 43, 44, 46, 82, 88, 92, 100, 103
sodium, 22, 28, 81, 82, 84, 90, 92, 100, 103
soil, 93
soil pollution, 93
soils, 25, 31
solvents, 13
South Africa, 7, 21, 57
South Asia, 5
Southeast Asia, 31, 61, 63, 65, 68, 69, 75
soy, 49
soybean, 82, 83
Spain, 36, 44, 45
species, 22, 24, 28, 30, 36, 44, 61, 91
specificity, 91
spermidine, 15
spermine, 15, 19
sports, 6
Sprague-Dawley rats, 94, 95, 97, 98
SPSS, 50
squamous cell, 79
squamous cell carcinoma, 79
stability, 22, 29, 52, 75, 76, 79, 80
stages, 62, 69, 72
standards, 16
starch, 22, 28
statistical analysis, 50

steroid, 55
stomach, 76, 79, 80, 82, 84, 93, 95, 98, 99
storage, 22, 28
strain, 22, 29
strains, 21, 22, 29
strategies, 5, 6
stress, 22, 28
structuring, 1
students, 4
substances, 22, 28, 35, 38, 40, 47, 49, 87, 93, 99
sugar, 36, 89
sulfate, 22, 28
sunlight, 81
supercritical, 95, 98
supplements, 36, 44, 53, 59, 60
surgical, 63
surveillance, 8, 11, 16, 17
susceptibility, 27, 64, 70, 100, 103
symptoms, 2
synchronous, 100, 104
synergistic, 82, 84, 100
synergistic effect, 100
synthesis, 49, 52
systemic biology, 66
systomics, 65

T

tamoxifen, 53, 59
targets, 64, 70
taste, 35, 87, 89
telephone, 53
telomerase, 100, 103
temperature, 35, 39
terephthalic acid, 15, 19
terrorism, 4, 8
testicular cancer, 42
Texas, 26, 31
TGF, 53, 58, 65
Thai, 63, 69, 71
Thailand, 13, 18, 22, 25, 28, 31, 61, 62, 68, 69, 71, 76, 79, 81
therapeutics, 56
therapy, 52, 55, 56, 64, 70

Index

Thomson, 8, 59
threat, 67
thymine, 15, 19
thyroid, 42
thyroid cancer, 42
tissue, 14, 36, 44, 52, 90, 91
Title III, 8
tobacco, 39, 82, 95, 98, 100, 103, 104
tobacco smoke, 39
tobacco smoking, 82, 100, 103
tomato, 22, 29
toxic, 12, 26, 40, 41, 88, 89, 93
toxic effect, 41
toxic substances, 40, 47, 93
toxicity, 12, 14, 26, 40, 41, 45, 46, 78, 81, 89, 94, 95, 97, 98
toxicological, 78, 95
toxicology, 40, 46, 89
toxin, 4, 22, 24, 26, 28
toxins, 2
TP53, 27, 32
trade, 8
traffic, 39
trans, 53, 57, 58
transcriptional, 52
transfer, 36, 44
transformation, 42, 90, 92
transforming growth factor, 65
transgenic, 53, 60, 95, 98
transition, 57
translocation, 95, 97
transmission, 3, 62, 68, 69
transport, 22, 29
tropical areas, 67
Trp, 42
tsunami, 62, 68
tuberculosis, 51
tumor, 49, 53, 55, 58, 59, 67, 88, 90, 91, 92, 95, 98
tumor cells, 50, 53, 58
tumor growth, 53, 59
tumor progression, 49
tumors, 49, 53, 58, 95, 98
tumour, 15, 20, 52
type 2 diabetes, 56

tyrosine, 88, 91, 95, 98

U

UK, 7, 13, 18
ultrasonography, 27
ultrasound, 64
ultraviolet, 24
uncertainty, 44
undernutrition, 1
United States, 64, 70, 79
urban areas, 5
urbanization, 57, 93
uric acid, 14
urinary, 15, 19, 27, 32, 40, 46, 73, 76, 80, 82, 84, 89
urinary bladder, 15, 19, 73, 82, 84, 89
urinary bladder cancer, 15, 89
urine, 22, 28, 42
Uruguay, 82, 84
Utah, 8

V

validity, 46
vegetable oil, 36, 44
vegetables, 8
vehicles, 11
vinegar, 12
vinyl chloride, 100, 105
virus, 26, 82
virus infection, 82
vitamin C, 76
vitamin D, 49, 50, 51, 52, 55, 56
vitamin D receptor, 50, 56
vitamin E, 76
vitamins, 40, 45
vomiting, 75

W

war on terror, 8
water, 22, 28, 61, 62, 65, 69, 76, 79, 80, 81, 82, 84, 93, 94, 97

water supplies, 79
waterfowl, 26
water-soluble, 22, 28
wheat, 21, 36, 43
wine, 22, 29, 99, 100, 104
Wistar rats, 82, 84
women, 8, 36, 44, 53, 56, 59, 100, 103
woods, 36, 44
workers, 27, 33
World Health Organization WHO, 17, 36, 43

X

xenografts, 53, 60

Y

yeast, 22, 29

Z

Zea mays, 31